HSC Year 12

MATHEMATICS ADVANCED

SIMON MELI

SERIES EDITOR: ROBERT YEN

A+

• 2020 UPDATED SYLLABUS • 2020 UPDATED SYLLABUS

- topic exams of HSC-style questions
- practice HSC and mini-HSC exams
- worked solutions with expert comments
- HSC exam topic grids (2011–2020)

PRACTICE EXAMS

A+ HSC Mathematics Advanced Practice Exams
1st Edition
Simon Meli
ISBN 9780170459235

Publishers: Robert Yen, Kirstie Irwin
Project editor: Tanya Smith
Cover design: Nikita Bansal
Text design: Alba Design
Project designer: Nikita Bansal
Permissions researcher: Corrina Gilbert
Production controller: Karen Young
Typeset by: Nikki M Group Pty Ltd

For product information and technology assistance,
in Australia call **1300 790 853**;
in New Zealand call **0800 449 725**

For permission to use material from this text or product, please email **aust.permissions@cengage.com**

ISBN 978 0 17 045923 5

Cengage Learning Australia
Level 7, 80 Dorcas Street
South Melbourne, Victoria Australia 3205

Cengage Learning New Zealand
Unit 4B Rosedale Office Park
331 Rosedale Road, Albany, North Shore 0632, NZ

For learning solutions, visit **cengage.com.au**

Printed in China by 1010 Printing International Limited.
1 2 3 4 5 6 7 25 24 23 22 21

ABOUT THIS BOOK

Introducing *A+ HSC Year 12 Mathematics*, a new series of study guides designed to help students revise the topics of the new HSC maths courses and achieve success in their exams. *A+* is published by Cengage, the educational publisher of *Maths in Focus* and *New Century Maths*.

For each HSC maths course, Cengage has developed a STUDY NOTES book and a PRACTICE EXAMS book. These study guides have been written by experienced teachers who have taught the new courses, some of whom are involved in HSC exam marking and writing. This is the first study guide series to be published after the first HSC exams of the new courses in 2020, so it incorporates the latest changes to the syllabus and exam format.

This book, *A+ HSC Year 12 Mathematics Advanced Practice Exams,* contains topic exams and practice HSC exams (including some past HSC questions), both written and formatted in the style of the HSC exams, with spaces for students to write answers. Worked solutions are provided along with the author's expert comments and advice, including how each exam question is marked. An HSC exam topic grid (2011–2020) guides students to where and how each topic has been tested in past HSC exams.

Mathematics Advanced Year 12 topics

1. Graphing functions
2. Trigonometric functions
3. Differentiation
4. Integration
5. Series, investments, loans and annuities
6. Statistics and bivariate data
7. Probability distributions

This book contains:

- 7 topic exams: 1-hour mini-HSC exams on each topic + worked solutions
- 2 practice mini-HSC exams: 1-hour exams + worked solutions
- 2 practice HSC exams: full (3-hour) exams + worked solutions
- HSC exam reference sheet of formulas
- bonus: worked solutions to the 2020 HSC exam.

The companion A+ STUDY NOTES book contains topic summaries and graded practice questions, grouped into the same 7 broad topics, including for each topic a concept map, glossary and HSC exam topic grid.

Both books can be used for revision after a topic has been learned, as well as for preparation for the trial and HSC exams. Before you begin any questions, make sure you have a thorough understanding of the topic you will be undertaking.

CONTENTS

YEAR 12 COURSE OVERVIEW

GRAPHING FUNCTIONS

Functions

- Polynomial functions
- Reciprocal functions
- Absolute value functions
- Exponential functions
- Logarithmic functions
- Domain and range
- Horizontal and vertical asymptotes
- x- and y-intercepts
- Applications of functions

Transformation of functions

- Translations
- Dilations
- Reflections in the x- and y-axes
- $y = kf(a(x + b)) + c$
- Combined transformations
- Domain and range
- Asymptotes and intercepts
- Symmetry and discontinuities

Solving equations and inequalities graphically

- Solving equations graphically or counting the number of solutions
- Solving linear and quadratic inequalities graphically

TRIGONOMETRIC FUNCTIONS

Transformations of trigonometric functions

- Graphing $y = kf(a(x + b)) + c$, where $f(x) = \sin x$, $\cos x$ or $\tan x$
- Dilations
- Reflections in the x- and y-axes
- Translations
- Domain and range
- Amplitude and centre
- Period
- Phase

Trigonometric equations

- Graphical solution, including counting number of solutions
- Algebraic solution

Applications of trigonometric functions

- Practical problems involving periodic phenomena, such as tides and electric currents.

DIFFERENTIATION

Differentiation rules

$y = x^n$ $\qquad \frac{dy}{dx} = nx^{n-1}$

$y = [f(x)]^n$ $\qquad \frac{dy}{dx} = nf'(x)[f(x)]^{n-1}$

Product rule

$y = uv$ $\qquad \frac{dy}{dx} = u\frac{dv}{dx} + v\frac{du}{dx}$

Quotient rule

$y = \frac{u}{v}$ $\qquad \frac{dy}{dx} = \frac{v\frac{du}{dx} - u\frac{dv}{dx}}{v^2}$

Chain rule

$y = g(u)$ where $u = f(x)$ $\qquad \frac{dy}{dx} = \frac{dy}{du} \times \frac{du}{dx}$

The first derivative

- Stationary point: $\frac{dy}{dx} = 0$
- Increasing function: $\frac{dy}{dx} > 0$
- Decreasing function: $\frac{dy}{dx} < 0$

Stationary points

- A stationary point occurs where $\frac{dy}{dx} = 0$
- Maximum turning point if $\frac{d^2y}{dx^2} < 0$
- Minimum turning point if $\frac{d^2y}{dx^2} > 0$
- Horizontal point of inflection is where $\frac{dy}{dx} = \frac{d^2y}{dx^2} = 0$ and change in concavity occurs.

Derivatives of trigonometric functions

$y = \sin x$ $\qquad \frac{dy}{dx} = \cos x$

$y = \cos x$ $\qquad \frac{dy}{dx} = -\sin x$

$y = \tan x$ $\qquad \frac{dy}{dx} = \sec^2 x$

Derivatives of exponential and logarithmic functions

$y = e^x$ $\qquad \frac{dy}{dx} = e^x$

$y = e^{f(x)}$ $\qquad \frac{dy}{dx} = f'(x)e^{f(x)}$

$y = a^x$ $\qquad \frac{dy}{dx} = (\ln a)a^x$

$y = \ln x$ $\qquad \frac{dy}{dx} = \frac{1}{x}$

$y = \ln f(x)$ $\qquad \frac{dy}{dx} = \frac{f'(x)}{f(x)}$

$y = \log_a x$ $\qquad \frac{dy}{dx} = \frac{1}{x \ln a}$

The second derivative and concavity

- Concave up: $\frac{d^2y}{dx^2} > 0$
- Concave down: $\frac{d^2y}{dx^2} < 0$
- $\frac{dy}{dx} = 0$ for point of inflection and check either side for change in concavity.

Optimisation and motion problems

- Maximum and minimum problems
- Displacement, velocity, acceleration

INTEGRATION

Anti-differentiation

- Opposite of differentiation
- Anti-derivative, primitive
- The indefinite integral $\int f(x)\,dx$:

$$\int x^n dx = \frac{1}{n+1}x^{n+1} + c$$

- The reverse chain rule:

$$\int f'(x)[f(x)]^n dx = \frac{1}{n+1}[f(x)]^{n+1} + c$$

Exponential functions

$$\int e^x dx = e^x + c$$

$$\int e^{ax+b} dx = \frac{1}{a}e^{ax+b} + c$$

$$\int a^x dx = \frac{1}{\ln a}a^x + c$$

The trapezoidal rule

$$\int_a^b f(x)\,dx \approx \frac{b-a}{2n}\{f(a) + f(b) + 2[f(x_1) + \cdots + f(x_{n-1})]\}$$

Areas between curves

$$A = \int_a^b [f(x) - g(x)]\,dx$$

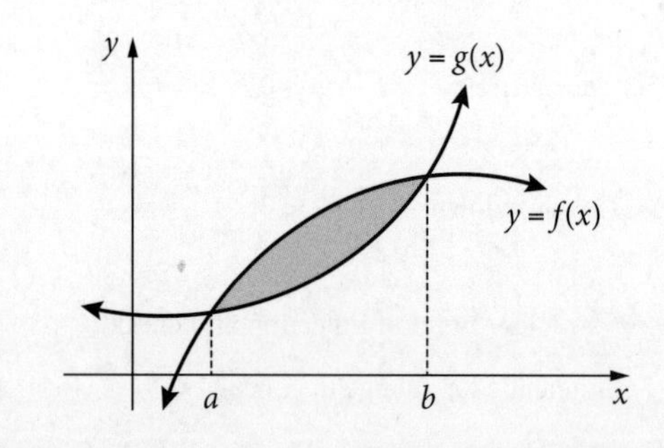

Trigonometric functions

$$\int \sin(ax+b)dx = -\frac{1}{a}\cos(ax+b) + c$$

$$\int \cos(ax+b)dx = \frac{1}{a}\sin(ax+b) + c$$

$$\int \sec^2(ax+b)dx = \frac{1}{a}\tan(ax+b) + c$$

Logarithmic functions

$$\int \frac{1}{x}\,dx = \ln|x| + c$$

$$\int \frac{f'(x)}{f(x)}\,dx = \ln|f(x)| + c$$

Area under a curve

The definite integral $\int_a^b f(x)\,dx$

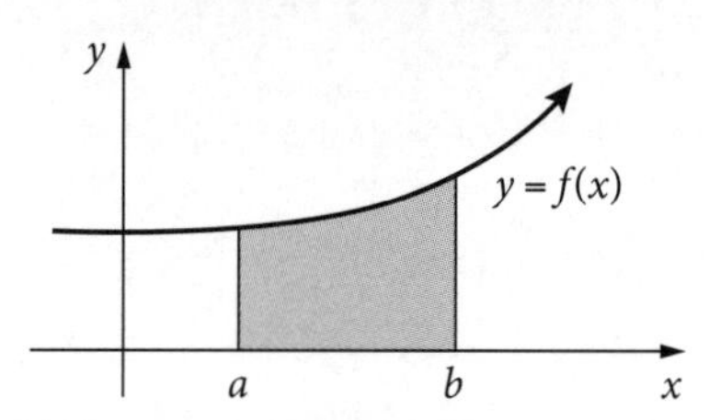

Applications of integration

- Given $f'(x)$ and an initial condition $f(a) = b$, find $f(x)$.
- Problems involving displacement, velocity, acceleration and rates of change.

SERIES, INVESTMENTS, LOANS AND ANNUITIES

Common content with the Mathematics Standard 2 course.

Arithmetic sequences and series

$T_n = a + (n - 1)d$

$T_n = T_{n-1} + d$

$S_n = \frac{n}{2}[2a + (n - 1)d]$

$S_n = \frac{n}{2}(a + l)$

Investments

- Compound interest
- Effective interest rate

Annuities

- Future value
- Present value
- By tables and recurrence relations
- By geometric series

Geometric sequences and series

$T_n = ar^{n-1}$

$T_n = rT_{n-1}$

$S_n = \frac{a(r^n - 1)}{r - 1}$

or $S_n = \frac{a(1 - r^n)}{1 - r}$

$S_\infty = \frac{a}{1 - r}, \quad |r| < 1$

Reducing balance loans

- By tables and recurrence relations
- By geometric series

STATISTICS AND BIVARIATE DATA

Other than variance, this entire topic is common content with the Mathematics Standard 2 course.

Types of data

- Categorical: nominal/ordinal
- Numerical: discrete/continuous

Measures of spread

- Range
- Quartiles, deciles and percentiles
- Interquartile range (IQR) = $Q_3 - Q_1$
- Variance, σ^2
- Standard deviation, σ
- Sample standard deviation, s
- An outlier is below $Q_1 - 1.5 \times \text{IQR}$ or above $Q_3 + 1.5 \times \text{IQR}$

Scatterplots

Used to graph bivariate data (2 variables)

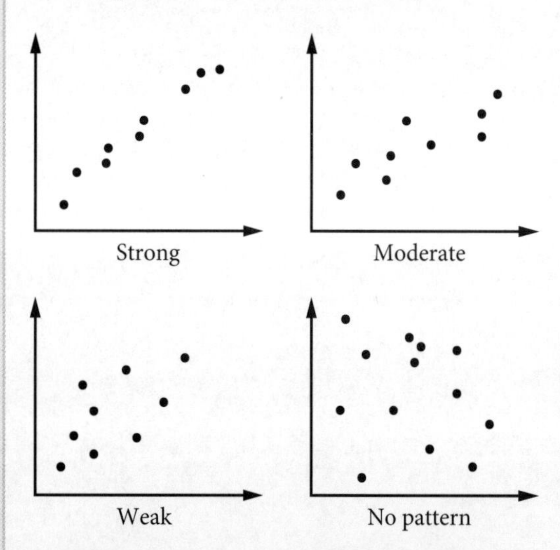

Line of best fit

- Dependent and independent variables
- Drawing by eye
- Using technology: least-squares regression line
- Interpolation and extrapolation

Measures of central tendency

- Mean, $\bar{x} = \dfrac{\text{sum of values}}{\text{number of values}}$

  $\bar{x} = \dfrac{\Sigma fx}{\Sigma f}$ for fx table
- Median: middle value
- Mode: most common value

The shape of a statistical distribution

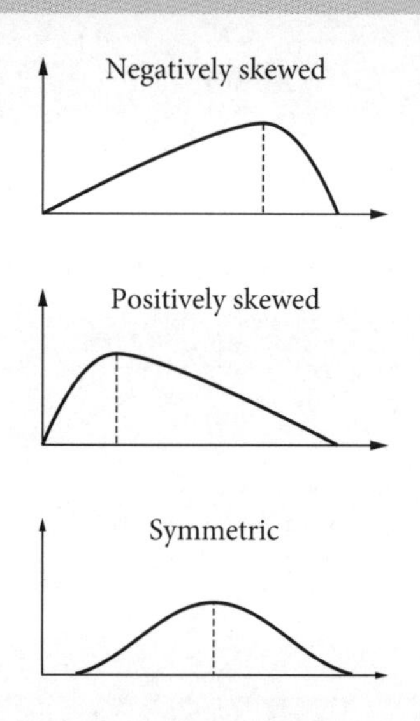

Correlation

- Pearson's correlation coefficient, r, where $-1 \le r \le 1$ for linear relationships:

$r = -1$:	strong, negative
$r = -0.5$:	moderate, negative
$r = 0$:	no correlation
$r = 0.5$:	moderate, positive
$r = 1$:	strong, positive

Common content with the Mathematics Standard 2 course.

PROBABILITY DISTRIBUTIONS

Continuous probability distributions

Probability density function (PDF)

$$\int_{-\infty}^{\infty} f(x)dx = 1$$

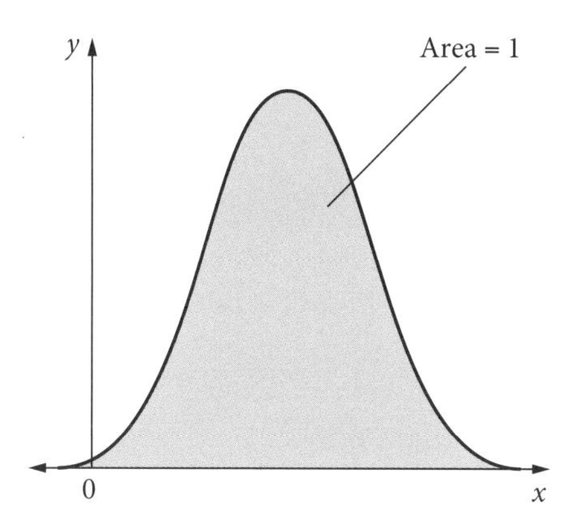

$$P(a \le X \le b) = \int_a^b f(x)dx$$

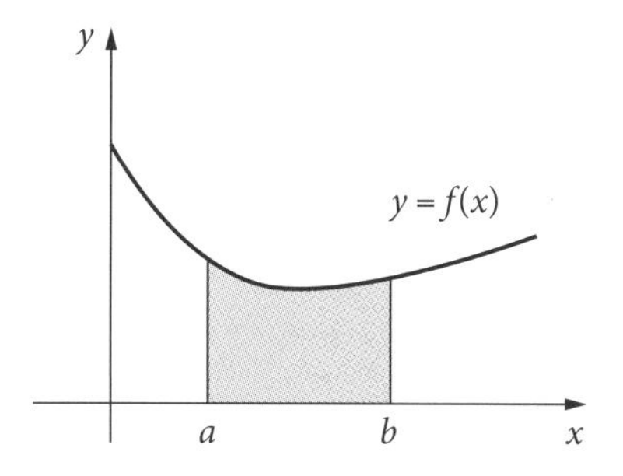

Cumulative distribution function (CDF)

$$F(x) = \int_a^x f(x)\,dx$$

Quantiles

- Median: $\int_a^x f(x)dx = \frac{1}{2}$
- Lower quartile: $\int_a^x f(x)dx = \frac{1}{4}$
- Upper quartile: $\int_a^x f(x)dx = \frac{3}{4}$
- Deciles: 3rd decile > bottom 30% of values
  → $\int_a^x f(x)dx = 0.3$
- Percentiles: 64th percentile > bottom 64% of values
  → $\int_a^x f(x)dx = 0.64$

Normal distribution

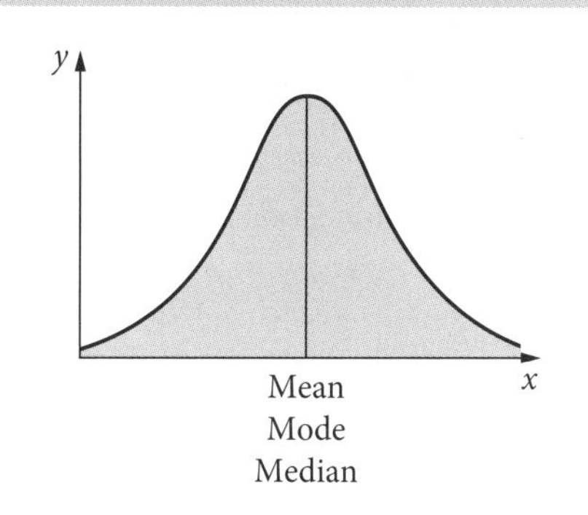

z-scores

- Measures number of standard deviations from the mean: $z = -1.6$ means 1.6 standard deviations below the mean.

$$z = \frac{x - \mu}{\sigma}$$

- **Empirical rule**

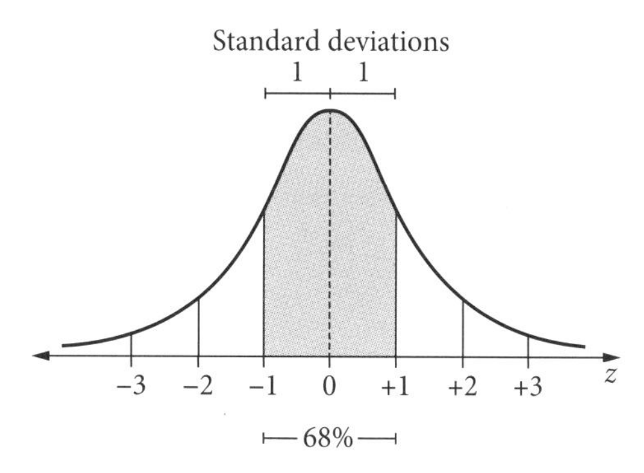

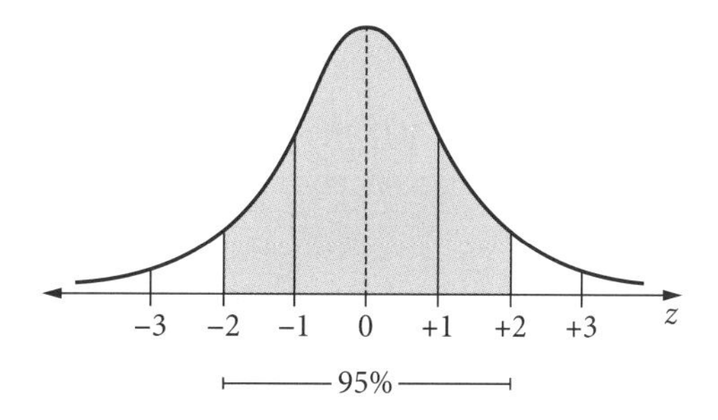

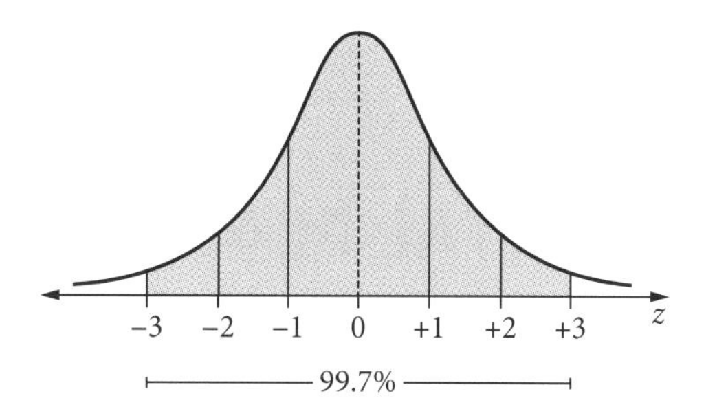

- Probability tables for z-scores
- Comparing z-scores

SYLLABUS REFERENCE GRID

Topic and subtopics	***A+ HSC Year 12 Mathematics Advanced Practice Exams* chapter**
FUNCTIONS	
MA-F2 Graphing techniques	1 Graphing functions
TRIGONOMETRIC FUNCTIONS	
MA-T3 Trigonometric functions and graphs	2 Trigonometric functions
CALCULUS	
MA-C2 Differential calculus C2.1 Differentiation of trigonometric, exponential and logarithmic functions C2.2 Rules of differentiation	3 Differentiation
MA-C3 Applications of differentiation C3.1 The first and second derivatives C3.2 Applications of the derivative	3 Differentiation
MA-C4 Integral calculus C4.1 The anti-derivative C4.2 Areas and the definite integral	4 Integration
FINANCIAL MATHEMATICS	
MA-M1 Modelling financial situations M1.1 Modelling investments and loans M1.2 Arithmetic sequences and series M1.3 Geometric sequences and series M1.4 Financial applications of sequences and series	5 Series, investments, loans and annuities
STATISTICAL ANALYSIS	
MA-S2 Descriptive statistics and bivariate data analysis S2.1 Data (grouped and ungrouped) and summary statistics S2.2 Bivariate data analysis	6 Statistics and bivariate data
MS-S3 Random variables S3.1 Continuous random variables S3.2 The normal distribution	6 Statistics and bivariate data 7 Probability distributions

ABOUT THE AUTHOR

Simon Meli teaches at Knox Grammar School, Wahroonga, and has taught mathematics for over 30 years. He has been involved in HSC marking and judging, and is a contributor to MANSW's annual HSC mathematics exam solutions.

A+ DIGITAL FLASHCARDS

Revise key terms and concepts online with the A+ Flashcards. Each topic for this course has a deck of digital flashcards you can use to test your understanding and recall. Just scan the QR code or type the URL into your browser to access them.

Note: You will need to create a free NelsonNet account.

https://get.ga/a-hsc-maths-advanced

HSC EXAM FORMAT

Mathematics Advanced HSC exam

The following information about the Mathematics Advanced HSC exam was correct at the time of printing in 2021. Please check the NESA website in case it has changed.
Visit www.educationstandards.nsw.edu.au, select 'Year 11 – Year 12', 'Syllabuses A–Z', 'Mathematics Advanced', then 'Assessment and reporting in Mathematics Advanced Stage 6'. Scroll down to the heading 'HSC examination specifications'.

	Questions	Marks	Recommended time
Section I	10 multiple-choice questions Mark answers on the multiple-choice answer sheet.	10	15 min
Section II	Approx. 21 short-answer questions, including 2 or more questions worth 4 or 5 marks. Write answers on the lines provided on the paper.	90	2 h 45 min
Total		100	3 h

Exam information and tips

- Reading time: 10 minutes; use this time to preview the whole exam.
- Working time: 3 hours
- Questions focus on Year 12 outcomes but Year 11 knowledge may be examined.
- Answers are to be written on the question paper.
- A reference sheet is provided at the back of the exam paper containing formulas.
- Common questions with the Mathematics Standard 2 HSC exam: 20–25 marks
- The 4- and 5-mark questions are usually complex problems that require many steps of working and careful planning.
- To help you plan your time, the halfway point of Section II is marked by a notice at the bottom of the relevant page; for example, 'Questions 11–23 are worth 46 marks in total'.
- Having 3 hours for a total of 100 marks means that you have an average of 1.8 minutes per mark (or 5 minutes for 3 marks).
- If you budget 15 minutes for Section I and 1 hour 15 minutes for each half of Section II, then you will have 15 minutes at the end of the exam to check over your work and complete questions you missed.

STUDY AND EXAM ADVICE

A journey of a thousand miles begins with a single step.

Lao Tzu (c. 570–490 BCE), Chinese philosopher

I've always believed that if you put in the work, the results will come.

Michael Jordan (1963–), American basketball player

Four PRACtical steps for maths study

1. Practise your maths

- Do your homework.
- Learning maths is about mastering a collection of skills.
- You become successful at maths by doing it more, through regular practice and learning.
- Aim to achieve a high level of understanding.

2. Rewrite your maths

- Homework and study are not the same thing. Study is your private 'revision' work for strengthening your understanding of a subject.
- Before you begin any questions, make sure you have a thorough understanding of the topic.
- Take ownership of your maths. Rewrite the theory and examples in your own words.
- Summarise each topic to see the 'whole picture' and know it all.

3. Attack your maths

- All maths knowledge is interconnected. If you don't understand one topic fully, then you may have trouble learning another topic.
- Mathematics is not an HSC course you can learn 'by halves' – you have to know it all!
- Fill in any gaps in your mathematical knowledge to see the 'whole picture'.
- Identify your areas of weakness and work on them.
- Spend most of your study time on the topics you find difficult.

4. Check your maths

- After you have mastered a maths skill, such as graphing a quadratic equation, no further learning or reading is needed, just more practice.
- Compared to other subjects, the types of questions asked in maths exams are conventional and predictable.
- Test your understanding with revision exercises, practice papers and past exam papers.
- Develop your exam technique and problem-solving skills.
- Go back to steps 1–3 to improve your study habits.

Topic summaries and concept maps

Summarise each topic when you have completed it, to create useful study notes for revising the course, especially before exams. Use a notebook or folder to list the important ideas, formulas, terminology and skills for each topic. This book is a good study guide, but educational research shows that effective learning takes place when you rewrite learned knowledge in your own words.

A good topic summary runs for 2 to 4 pages. It is a condensed, personalised version of your course notes. This is your interpretation of a topic, so include your own comments, symbols, diagrams, observations and reminders. Highlight important facts using boxes and include a glossary of key words and phrases.

A concept map or mind map is a topic summary in graphic form, with boxes, branches and arrows showing the connections between the main ideas of the topic. This book contains examples of concept maps. The topic name is central to the map, with key concepts or subheadings listing important details and formulas. Concept maps are powerful because they present an overview of a topic on one large sheet of paper. Visual learners absorb and recall information better using concept maps.

When compiling a topic summary, use your class notes, your textbook and the A+ Study Notes book that accompanies this book. Ask your teacher for a copy of the course syllabus or the school's teaching program, which includes the objectives and outcomes of every topic in dot point form.

Attacking your weak areas

Most of your study time should be spent on attacking your weak areas to fill in any gaps in your maths knowledge. Don't spend too much time on work you already know well, unless you need a confidence boost! Ask your teacher, use this book or your textbook to improve the understanding of your weak areas and to practise maths skills. Use your topic summaries for general revision, but spend longer study periods on overcoming any difficulties in your mastery of the course.

Practising with past exam papers

Why is practising with past exam papers such an effective study technique? It allows you to become familiar with the format, style and level of difficulty expected in an HSC exam, as well as the common topic areas tested. Knowing what to expect helps alleviate exam anxiety. Remember, mathematics is a subject in which the exam questions are fairly predictable. The exam writers are not going to ask too many unusual questions. By the time you have worked through many past exam papers, this year's HSC paper won't seem that much different.

Don't throw your old exam papers away. Use them to identify your mistakes and weak areas for further study. Revising topics and then working on mixed questions is a great way to study maths. You might like to complete a past HSC exam paper under timed conditions to improve your exam technique.

Past HSC exam papers are available at the NESA website: visit www.educationstandards.nsw.edu.au and select 'Year 11 – Year 12', 'HSC exam papers'. NESA marking feedback and guidelines can also be viewed there. You can find past HSC exam papers with solutions online, in bookstores, at the Mathematical Association of NSW (www.mansw.nsw.edu.au) and at your school (ask your teacher) or library.

Preparing for an exam

- Make a study plan early; don't leave it until the last minute.
- Read and revise your topic summaries.
- Work on your weak areas and learn from your mistakes.
- Don't spend too much time studying work you know already.
- Revise by completing revision exercises and past exam papers or assignments.
- Vary the way you study so that you don't become bored: ask someone to quiz you, voice-record your summary, design a poster or concept map, or explain the work to someone.
- Anticipate the exam:
  - How many questions will there be?
  - What are the types of questions: multiple-choice, short-answer, long-answer, problem-solving?
  - Which topics will be tested?
  - How many marks are there in each section?
  - How long is the exam?
  - How much time should I spend on each question/section?
  - Which formulas are on the reference sheet and how do I use them in the exam?

During an exam

1. Bring all of your equipment, including a ruler and calculator (check that your calculator works and is in RADIANS mode for trigonometric functions and DEGREES for trigonometric measurements). A highlighter pen may help for tables, graphs and diagrams.
2. Don't worry if you feel nervous before an exam – this is normal and helps you to perform better; however, being too casual or too anxious can harm your performance. Just before the exam begins, take deep, slow breaths to reduce any stress.
3. Write clearly and neatly in black or blue pen, not red. Use a pencil only for diagrams and constructions.
4. Use the **reading time** to browse through the exam to see the work that is ahead of you and the marks allocated to each question. Doing this will ensure you won't miss any questions or pages. Note the harder questions and allow more time to work on them. Leave them if you get stuck, and come back to them later.
5. Attempt every question. It is better to do most of every question and score some marks, rather than ignore questions completely and score 0 for them. Don't leave multiple-choice questions unanswered! Even if you guess, you have a chance of being correct.
6. Easier questions are usually at the beginning, with harder ones at the end. Do an easy question first to boost your confidence. Some students like to leave multiple-choice questions until last so that, if they run out of time, they can make quick guesses. However, some multiple-choice questions can be quite difficult.
7. Read each question and identify what needs to be found and what topic/skill it is testing. The number of marks indicates how much time and working out is required. Highlight any important keywords or clues. Do you need to use the answer to the previous part of the question?
8. After reading each question, and before you start writing, spend a few moments planning and thinking.
9. You don't need to be writing all of the time. What you are writing may be wrong and a waste of time. Spend some time considering the best approach.

10. Make sure each answer seems reasonable and realistic, especially if it involves money or measurement.
11. Show all necessary working, write clearly, draw big diagrams, and set your working out neatly. Write solutions to each part underneath the previous step so that your working out goes down the page, not across.
12. Use a ruler to draw (or read) half-page graphs with labels and axes marked, or to measure scale diagrams.
13. Don't spend too much time on one question. Keep an eye on the time.
14. Make sure you have answered the question. Did you remember to round the answer and/or include units? Did you use all of the relevant information given?
15. If a hard question is taking too long, don't get bogged down. If you're getting nowhere, retrace your steps, start again, or skip the question (circle it) and return to it later with a clearer mind.
16. If you make a mistake, cross it out with a neat line. Don't scribble over it completely or use correction fluid or tape (which is time-consuming and messy). You may still score marks for crossed-out work if it is correct, but don't leave multiple answers! Keep track of your answer booklets and ask for more writing paper if needed.
17. Don't cross out or change an answer too quickly. Research shows that often your first answer is the correct one.
18. Don't round your answer in the middle of a calculation. Round at the end only.
19. Be prepared to write words and sentences in your answers, but don't use abbreviations that you've just made up. Use correct terminology and write 1 or 2 sentences for 2 or 3 marks, not mini-essays.
20. If you have time at the end of the exam, double-check your answers, especially for the more difficult or uncertain questions.

Ten exam habits of the best HSC students

1. Has clear and careful working and checks their answers
2. Has a strong understanding of basic algebra and calculation
3. Reads (and answers) the whole question
4. Chooses the simplest and quickest method
5. Checks that their answer makes sense or sounds reasonable
6. Draws big, clear diagrams with details and labels
7. Uses a ruler for drawing, measuring and reading graphs
8. Can explain answers in words when needed, in 1–2 clear sentences
9. Uses the previous parts of a question to solve the next part of the question
10. Rounds answers at the end, not before

Further resources

Visit the NESA website (www.educationstandards.nsw.edu.au) for the following resources. Select 'Year 11 – Year 12' and then 'Syllabuses A–Z' or 'HSC exam papers'.

- Mathematics Advanced Syllabus
- Past HSC exam papers, including marking feedback and guidelines
- Sample HSC questions/exam papers and marking guidelines

Before 2019, 'Mathematics Advanced' was called 'Mathematics'. For these exam papers, select 'Year 11 – Year 12', 'Resources archive', 'HSC exam papers archive'.

MATHEMATICAL VERBS

A glossary of 'doing words' common in maths problems and HSC exams

analyse
study in detail the parts of a situation

apply
use knowledge or a procedure in a given situation

calculate
See **evaluate**

classify/identify
state the type, name or feature of an item or situation

comment
express an observation or opinion about a result

compare
show how two or more things are similar or different

complete
fill in detail to make a statement, diagram or table correct or finished

construct
draw an accurate diagram

convert
change from one form to another, for example, from a fraction to a decimal, or from kilograms to grams

decrease
make smaller

describe
state the features of a situation

estimate
make an educated guess for a number, measurement or solution, to find roughly or approximately

evaluate/calculate
find the value of a numerical expression, for example, 3×8^2 or $4x + 1$ when $x = 5$

expand
remove brackets in an algebraic expression, for example, expanding $3(2y + 1)$ gives $6y + 3$

explain
describe why or how

give reasons
show the rules or thinking used when solving a problem. *See also* **justify**

graph
display on a number line, number plane or statistical graph

hence find/prove
calculate an answer or prove a result using previous answers or information supplied

identify
See **classify**

increase
make larger

interpret
find meaning in a mathematical result

justify
give reasons or evidence to support your argument or conclusion. *See also* **give reasons**

measure
determine the size of something, for example, using a ruler to determine the length of a pen

prove
See **show/prove that**

recall
remember and state

show/prove that
(in questions where the answer is given) use calculation, procedure or reasoning to prove that an answer or result is true

simplify
express a result such as a ratio or algebraic expression in its most basic, shortest, neatest form

sketch
draw a rough diagram that shows the general shape or ideas (less accurate than **construct**)

solve
calculate the value(s) of an unknown pronumeral in an equation or inequality

state
See **write**

substitute
replace part of an expression with another, equivalent expression

verify
check that a solution or result is correct, usually by substituting back into an equation or referring back to the problem

write/state
give an answer, formula or result without showing any working or explanation (This usually means that the answer can be found mentally, or in one step)

SYMBOLS AND ABBREVIATIONS

$=$	is equal to
$\neq$	is not equal to
$\approx$	is approximately equal to
$<$	is less than
$>$	is greater than
$\leq$	is less than or equal to
$\geq$	is greater than or equal to
$(\ )$	parentheses, round brackets
$[\ ]$	(square) brackets
$\{\ \}$	braces
$\pm$	plus or minus
π	pi = 3.141 59...
$\equiv$	is congruent/identical to
$0.\dot{1}5\dot{2}$	the recurring decimal 0.152 152...
$^\circ$	degree
$\angle$	angle
Δ	triangle
$\parallel$	is parallel to
$\perp$	is perpendicular to
$\therefore$	therefore
$\sqrt{\ }$	square root
$\sqrt[3]{\ }$	cube root
∞	infinity
$\lvert x \rvert$	absolute value of x
$\lim\limits_{h \to 0}$	the limit as $h \to 0$
$\frac{dy}{dx}, y', f'(x)$	the 1st derivative of $y, f(x)$
$\frac{d^2y}{dx^2}, y'', f''(x)$	the 2nd derivative of $y, f(x)$
$\int f(x)\,dx$	the integral of $f(x)$

$[a, b], a \leq x \leq b$	the interval of x-values from a to b (including a and b)
$(a, b), a < x < b$	the interval of x-values between a and b (excluding a and b)
S 37° W	a compass bearing
217°	a true bearing
$P(E)$	the probability of event E occurring
$P(\bar{E})$	the probability of event E not occurring
$P(A \mid B)$	the probability of A given B
$A \cup B$	A union B, A or B
$A \cap B$	A intersection B, A and B
PDF	probability density function
CDF	cumulative distribution function
LHS	left-hand side
RHS	right-hand side
p.a.	per annum (per year)
$\bar{x}$	the mean
$\mu = E(X)$	the population mean, expected value
σ_n	the standard deviation
$\text{Var}(X) = \sigma^2$	the variance
Σ	the sum of, sigma
Q_1	first quartile or lower quartile
Q_2	median (second quartile)
Q_3	third quartile or upper quartile
IQR	interquartile range
α	alpha
θ	theta
m	gradient

A+ HSC YEAR 12 MATHEMATICS

STUDY NOTES

Authors:

Tania Eastcott
Rachel Eastcott

Sarah Hamper

Karen Man
Ashleigh Della Marta

Jim Green
Janet Hunter

PRACTICE EXAMS

Authors:

Adrian Kruse

Simon Meli

John Drake

Jim Green
Janet Hunter

CHAPTER 1
TOPIC EXAM

Graphing functions

MA-F2 Graphing techniques

- A reference sheet is provided on page 195 at the back of this book
- For questions in Section II, show relevant mathematical reasoning and/or calculations

Section I – 3 questions, 3 marks
- Attempt Questions 1–3
- Allow about 5 minutes for this section

Section II – 8 questions, 30 marks
- Attempt Questions 4–11
- Allow about 55 minutes for this section

Reading time: 4 minutes
Working time: 1 hour
Total marks: 33

Section I

- Attempt Questions 1–3
- Allow about 5 minutes for this section

3 marks

Question 1

Given $f(x) = x^3 + x^2 - x$, what is the value of $f(-2)$?

A -14

B -10

C -6

D -2

Question 2

What is the gradient of the line with equation $2x - 5y + 1 = 0$?

A $-\frac{5}{2}$

B $-\frac{2}{5}$

C $\frac{2}{5}$

D $\frac{5}{2}$

Question 3

A relation is defined by the set of points below.

$$\{(0,5), (-2,1), (0,3), (-4,1)\}$$

Which term best describes this relation?

A One-to-one

B Many-to-many

C One-to-many

D Many-to-one

Section II

- Attempt Questions 4–11 **30 marks**
- Allow about 55 minutes for this section
- Answer the questions in the spaces provided. These spaces provide guidance for the expected length of response.
- Your responses should include relevant mathematical reasoning and/or calculations.

Question 4 (2 marks)

Express $\frac{2}{\sqrt{5}-1}$ with a rational denominator. 2 marks

Question 5 (4 marks)

The diagram shows the graph of $f(x) = 2x^2 - 6x - 1$.

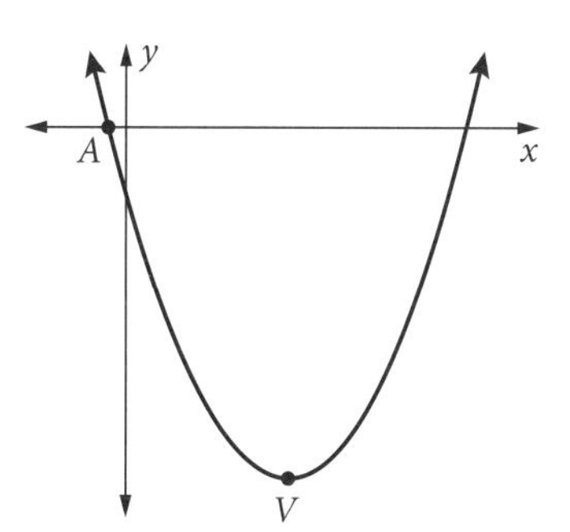

a Find the coordinates of the vertex, V. 2 marks

b Find the x-coordinate of the intercept marked A. 2 marks

Question 6 (2 marks)

Simplify $\dfrac{1}{n-1} - \dfrac{2}{n}$. 2 marks

Question 7 (6 marks)

Jamjet Industries produces galvanised bolts for the roofing industry. Due to manufacturing constraints, the maximum number of bolts that can be produced in a month is 120 000.

The cost of production is \$50 000 plus 75 cents per bolt.

The bolts are sold for \$1.75 each and the graph that represents Jamjet's revenue, $\$R$, from selling n bolts appears on the axes below.

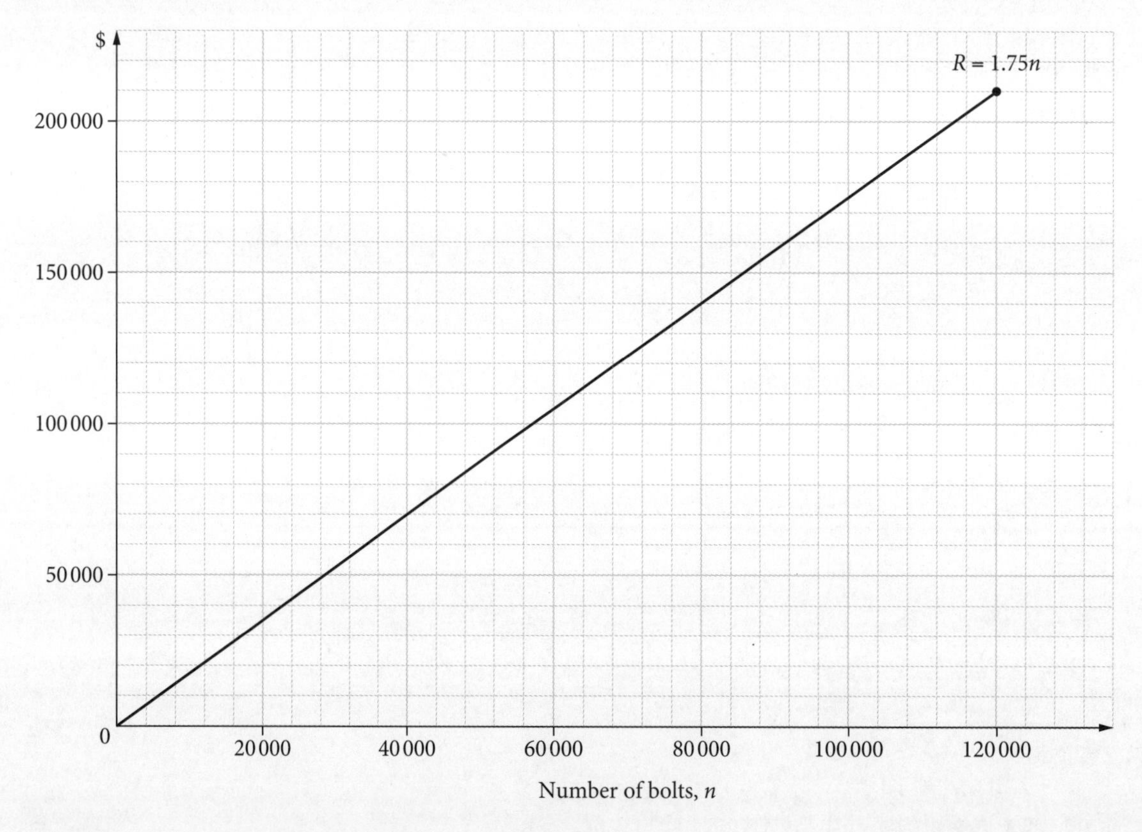

a Find the equation for the cost, $\$C$, of producing n bolts. 1 mark

b Graph the equation for $\$C$ on the axes above. 1 mark

c What is the practical significance of the gradient of the line you drew in part **b**? 1 mark

d By using the graphs drawn, or otherwise, find the number of bolts Jamjet must sell in a month to break even. 1 mark

e What is the maximum monthly profit the company can make? 2 marks

Questions 4–7 are worth 14 marks in total (Section II halfway point)

Question 8 (6 marks)

The graph of $y = f(x)$ appears below with x-intercepts at -6, -1 and 3, as shown.

There is a maximum turning point at $(-4, 5)$ and a minimum turning point at $(1.5, -3)$.

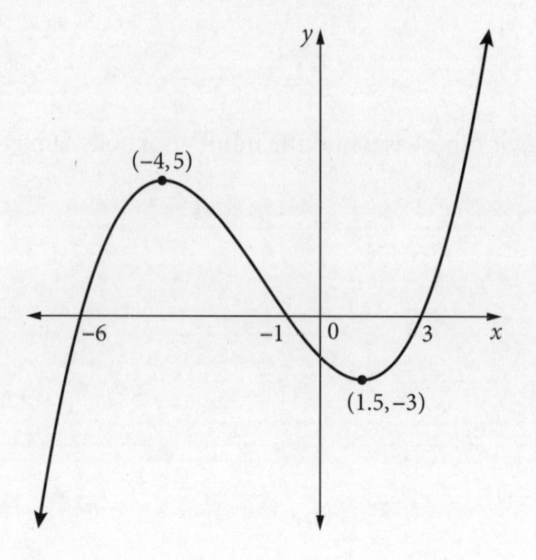

a Explain how you know that this is the graph of a function. 1 mark

b Solve $f(x) \leq 0$, giving the solution in interval notation. 2 marks

c For what value(s) of k does the equation $f(x) - k = 0$ have exactly one solution? 1 mark

d The graph of $y = f(x + a) + b$ has a maximum turning point at the origin.

Find the values of a and b. 2 marks

Question 9 (3 marks) ©NESA 2020 HSC EXAM, QUESTION 24

The circle $x^2 - 6x + y^2 + 4y - 3 = 0$ is reflected in the x-axis.

Sketch the reflected circle, showing the coordinates of the centre and the radius. 3 marks

Question 10 (5 marks)

An interval joining the points $A(-5,-4)$ and $B(-1,-3)$ is shown on the number plane below.

A line, l, is drawn through B so that it is perpendicular to AB and meets the x-axis at C and the y-axis at D.

Find the area of triangle COD. 5 marks

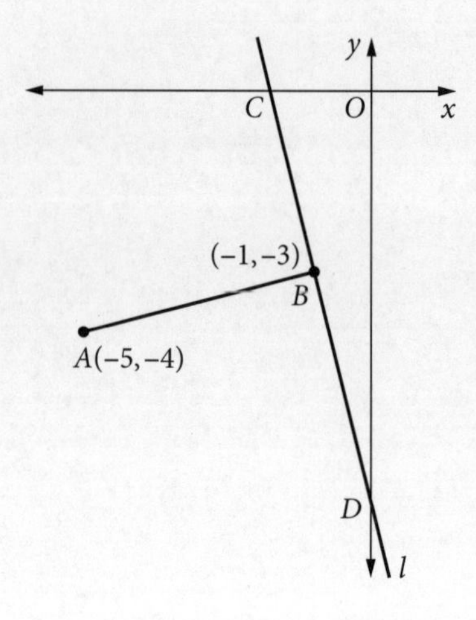

Question 11 (2 marks) ©NESA 2020 SAMPLE HSC EXAM, QUESTION 17

Given the function $f(x) = x^2 + 2$ and $g(x) = \sqrt{x-6}$, sketch $y = f(g(x))$ over its natural domain. 2 marks

END OF PAPER

WORKED SOLUTIONS

Section I (1 mark each)

Question 1

D $f(x) = (-2)^3 + (-2)^2 - (-2)$
$= -8 + 4 + 2$
$= -2$

Be careful when substituting negatives.

Question 2

C $2x - 5y + 1 = 0$
$5y = 2x + 1$
$y = \frac{2}{5}x + \frac{1}{5}$

Gradient is $\frac{2}{5}$.

Common straightforward question.

Question 3

B The x-value 0 produces 2 values of y (namely 5 and 3).

The y-value 1 produces 2 values of x (namely -2 and -4).

So this relation is many-to-many.

This is new to the course. Make sure you know the relations terminology.

Section II (✓ = 1 mark)

Question 4 (2 marks)

$$\frac{2}{\sqrt{5}-1} = \frac{2}{\sqrt{5}-1} \times \frac{\sqrt{5}+1}{\sqrt{5}+1} \checkmark$$
$$= \frac{2(\sqrt{5}+1)}{(\sqrt{5})^2 - 1^2}$$
$$= \frac{2(\sqrt{5}+1)}{4}$$
$$= \frac{\sqrt{5}+1}{2} \checkmark$$

Straightforward but appears quite regularly in HSC exams.

Question 5 (4 marks)

a The vertex lies on the axis of symmetry which is at $x = -\frac{b}{2a} = -\frac{-6}{2 \times 2} = \frac{3}{2}$. ✓

For the y-coordinate, find

$$f\left(\frac{3}{2}\right) = 2\left(\frac{3}{2}\right)^2 - 6\left(\frac{3}{2}\right) - 1$$
$$= \frac{9}{2} - 9 - 1$$
$$= -5\frac{1}{2} \checkmark$$

The vertex is $\left(\frac{3}{2}, -5\frac{1}{2}\right)$.

Straightforward question. Make sure your answer makes sense when compared to the diagram.

b Solve $f(x) = 0$ to find the x-intercepts.

$2x^2 - 6x - 1 = 0$

$$\text{So } x = \frac{-b \pm \sqrt{b^2 - 4ac}}{2a}$$
$$= \frac{6 \pm \sqrt{(-6)^2 - (4 \times 2 \times -1)}}{2 \times 2}$$
$$= \frac{3 \pm \sqrt{11}}{2} \checkmark$$

But the x coordinate of A is negative,

so $x = \frac{3 - \sqrt{11}}{2}$. ✓

Always keep an eye on the diagram to make sure your answer makes sense.

Question 6 (2 marks)

$$\frac{1}{n-1} - \frac{2}{n} = \frac{n}{n(n-1)} - \frac{2(n-1)}{n(n-1)} \checkmark$$
$$= \frac{n - 2(n-1)}{n(n-1)}$$
$$= \frac{2 - n}{n(n-1)} \checkmark$$

Take care when expanding with negatives.

Question 7 (6 marks)

a Cost = \$50 000 + \$0.75 per bolt

$C = 50\,000 + 0.75n$ ✓

All of Question 7 is common content with Maths Standard 2.

b

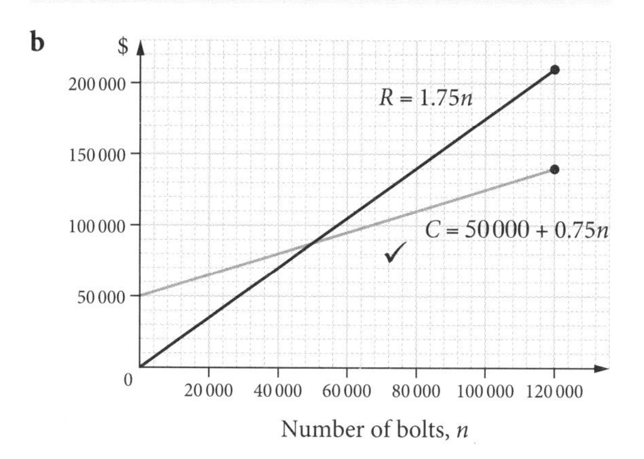

c The gradient, 0.75, represents the cost of production per bolt. ✓

Understanding the practical significance is key.

d From the graph, the point of intersection occurs at $n = 50\,000$. So Jamjet must sell 50 000 bolts in a month to break even. ✓

Can also be solved algebraically by simultaneous equations.

e The maximum profit is when the difference between the graphs of R and C is the greatest. As the maximum number of bolts produced is 120 000, this will occur at $n = 120\,000$. ✓

When $n = 120\,000$,

$$R = 1.75 \times 120\,000$$
$$= 210\,000$$

$$C = 50\,000 + 0.75 \times 120\,000$$
$$= 140\,000$$

So maximum profit = \$70 000. ✓

The values for R and C can also be read from the graph.

Question 8 (6 marks)

a Each value of x corresponds to a unique value of y, so it would pass the vertical line test. ✓

Know the functions definitions.

b Look at the x-values of the parts of the graph below the x-axis.

$(-\infty,-6] \cup [-1,3]$ ✓✓

(✓ for just one of the solutions)

Learn to solve equations graphically, not just algebraically. Learn to use interval notation correctly (writing inequalities in bracket form).

c Solving $f(x) - k = 0$ is the same as $f(x) = k$.

We require the horizontal line $y = k$ to meet the graph only once. So the line needs to be above $y = 5$ or below $y = -3$.
So $k < -3$ or $k > 5$.

$(-\infty,-3) \cup (5,\infty)$ ✓

d Graph of $y = f(x)$ must be translated 4 units right and 5 units down: $y = f(x - 4) - 5$.

$a = -4$ ✓ and $b = -5$ ✓

Question 9 (3 marks)

$$x^2 - 6x + y^2 + 4y - 3 = 0$$
$$x^2 - 6x + 9 + y^2 + 4y + 4 = 3 + 9 + 4$$
$$(x - 3)^2 + (y + 2)^2 = 16 \checkmark$$

Centre is $(3,-2)$ and radius is $\sqrt{16} = 4$.

If the circle is reflected in the x-axis, the centre moves to $(3,2)$ but the radius remains unchanged. ✓

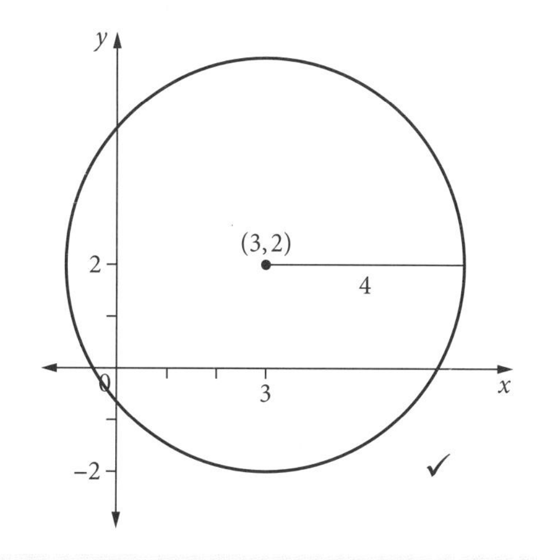

Complete the square to find the centre and radius. Sketch a clear diagram and use appropriate detail to describe the circle. Check that you have answered the question by including the graph.
A common error was incorrectly stating that the radius was 16 rather than 4.

Question 10 (5 marks)

$m_{AB} = \frac{-3-(-4)}{-1-(-5)} = \frac{1}{4}$ ✓

$m_l = -4$ ✓

$$\begin{aligned} y - y_1 &= m(x - x_1) \\ y + 3 &= -4(x+1) \\ &= -4x - 4 \\ y &= -4x - 7 \end{aligned}$$ ✓

So D is $(0, -7)$.

Set $y = 0$.

$$\begin{aligned} 0 &= -4x - 7 \\ 4x &= -7 \\ x &= -\frac{7}{4}, \text{ so } C \text{ is the point } \left(-\frac{7}{4}, 0\right). \end{aligned}$$

$OD = 7$ and $OC = \frac{7}{4}$ ✓

$$\begin{aligned} \text{Area} &= \frac{1}{2} \times \frac{7}{4} \times 7 \\ &= \frac{49}{8} \text{ units}^2 \end{aligned}$$ ✓

Straightforward 5-mark question but note that there is no scaffolding (instructions guiding you through the steps of answering it).

Question 11 (2 marks)

Domain of $g(x) = \sqrt{x-6}$ is $x \geq 6$.

This is the domain of $f(g(x))$.

$$\begin{aligned} f(g(x)) &= (\sqrt{x-6})^2 + 2 \\ &= x - 6 + 2 \\ &= x - 4 \quad (x \geq 6) \end{aligned}$$ ✓

Always check for restrictions to the domain.

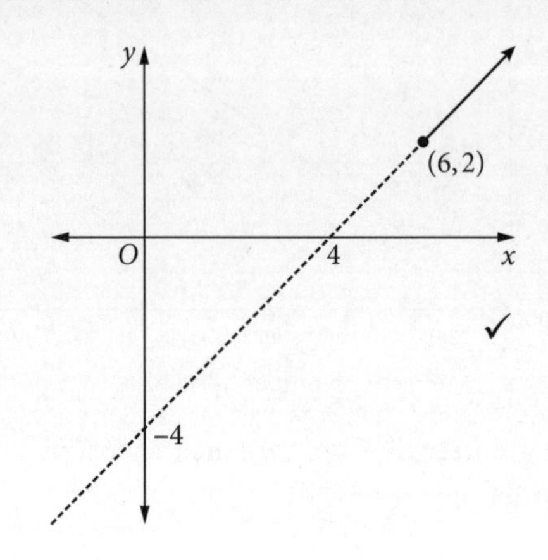

Straightforward line to graph, but only for $x \geq 6$.

HSC exam topic grid (2011–2020)

This table shows the coverage of this topic in past HSC exams by question number. The past exams can be downloaded from the NESA website (www.educationstandards.nsw.edu.au) by selecting 'Year 11 – Year 12', 'HSC exam papers'. NESA marking feedback and guidelines can also be found there.

Before 2020, 'Mathematics Advanced' was called 'Mathematics'. For these exams, select 'Year 11 – Year 12', 'Resources archive', 'HSC exam papers archive'.

	Graphing functions	Transformations of functions (introduced 2020)	Solving equations and inequalities graphically
2011			
2012			14(a)(iv)
2013	3, 16(b)		15(c)
2014		2	
2015	2, 12(d), 13(b)(i), 16(a)(i)–(ii)		8
2016	4, 11(a)	3	
2017	1, 11(h)	11(f)	
2018	3		
2019	12(a)		13(e)
2020 new course	1, 2	2, 5, **24**	11

Questions in **bold** can be found in this chapter.

CHAPTER 2
TOPIC EXAM

2

Trigonometric functions

MA-T3 Trigonometric functions and graphs

- A reference sheet is provided on page 195 at the back of this book
- For questions in Section II, show relevant mathematical reasoning and/or calculations

Reading time: 4 minutes
Working time: 1 hour
Total marks: 33

Section I – 3 questions, 3 marks
- Attempt Questions 1–3
- Allow about 5 minutes for this section

Section II – 6 questions, 30 marks
- Attempt Questions 4–11
- Allow about 55 minutes for this section

Section I

- Attempt Questions 1–3 **3 marks**
- Allow about 5 minutes for this section

Question 1

Which expression gives the value of x in this triangle?

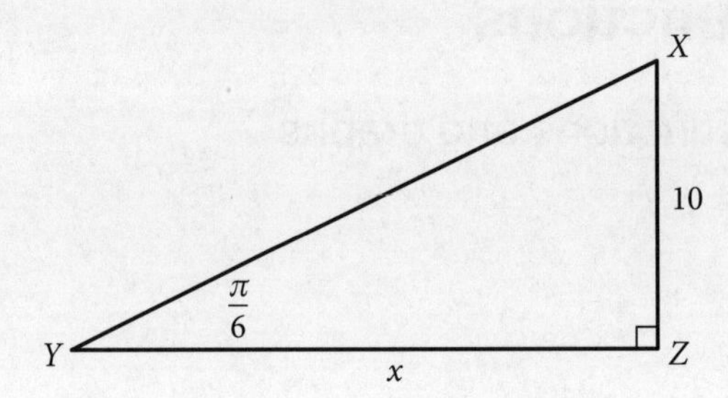

A $10\tan\left(\frac{\pi}{3}\right)$

B $10\sin\left(\frac{\pi}{3}\right)$

C $10\sin\left(\frac{\pi}{6}\right)$

D $10\tan\left(\frac{\pi}{6}\right)$

Question 2

Angela solves the equation $\sin^2\theta = \frac{2}{9}$ for $0° \le \theta \le 360°$.

How many solutions are there?

A 0

B 1

C 2

D 4

Question 3

Which expression has the same value as $\sin\left(\frac{4\pi}{3}\right)$?

A $\sin\left(\frac{\pi}{3}\right)$

B $-\sin\left(\frac{\pi}{3}\right)$

C $\cos\left(\frac{\pi}{3}\right)$

D $-\cos\left(\frac{\pi}{3}\right)$

9780170459235

Section II

- Attempt Questions 4–11 **30 marks**
- Allow about 55 minutes for this section
- Answer the questions in the spaces provided. These spaces provide guidance for the expected length of response.
- Your responses should include relevant mathematical reasoning and/or calculations.

Question 4 (3 marks)

In the right-angled triangle ABC, $\angle BAC = 30°$ and $AB = 15$ cm.

AD is drawn perpendicular to AC and is 10 cm long. CD is joined.

Find the exact length of CD. 3 marks

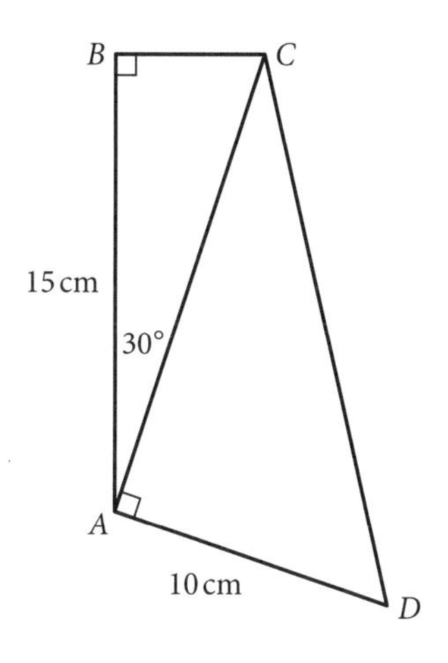

Question 5 (5 marks)

The sector below has centre O and a radius 12 cm. The arc AB subtends an angle of 40° at O.

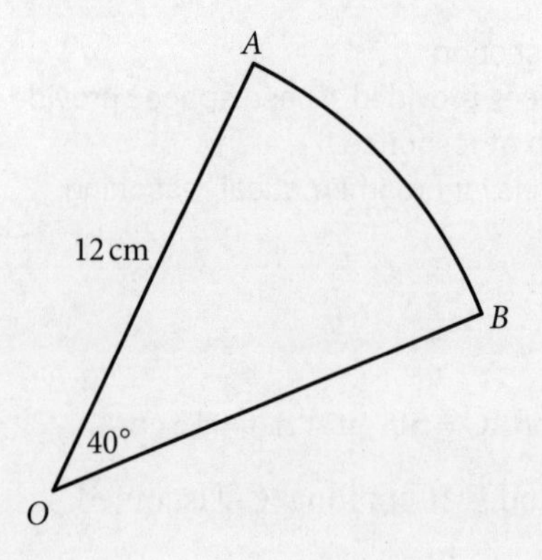

a Express the size of $\angle AOB$ in radians. 1 mark

b Find the exact perimeter of the sector. 2 marks

c Find the exact area of the sector. 2 marks

Question 6 (3 marks)

Find all possible values of θ, correct to the nearest degree. 3 marks

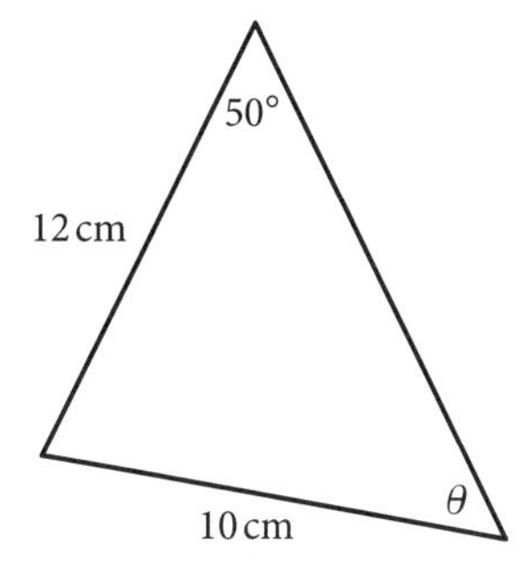

Question 7 (1 mark)

If $\cos\theta = \sin(2\theta + 15°)$, find θ. 1 mark

Question 8 (4 marks)

a Show that $\cot^2\theta\sin\theta + \sin\theta = \operatorname{cosec}\theta$ 2 marks

b Hence, or otherwise, solve 2 marks

$$2\cot^2\theta\sin\theta + 2\sin\theta = -4$$

for all values of θ in the domain $[-\pi, \pi]$.

Questions 4–8 are worth 16 marks in total (Section II halfway point)

Question 9 (5 marks)

A flagpole, BD, sits on flat level ground.

It is supported by two cables, AD and CD, as shown below. The cable AD is inclined at 30° to the horizontal and is tied 18 m from the foot of the pole.

The second cable is tied at point C, 12 m from A and on a bearing of 210° from A.

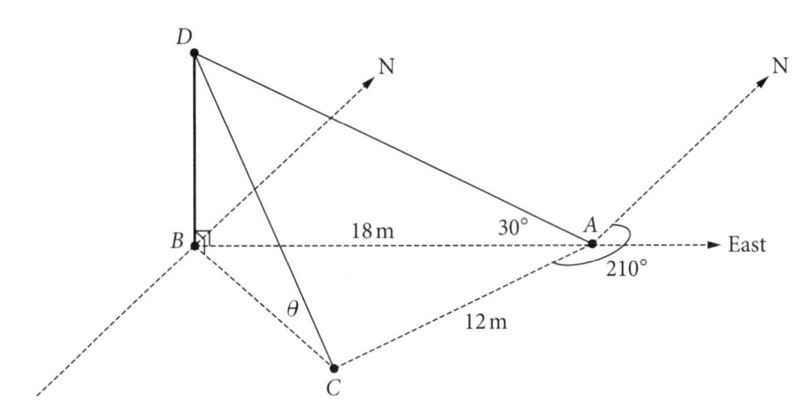

a Show that $\angle BAC = 60°$. 1 mark

b Find the exact height of the flagpole. 1 mark

c Find the exact distance BC. 2 marks

d Find θ, correct to the nearest degree. 1 mark

TOPIC EXAM

Question 10 (6 marks) ©NESA 2009 HSC EXAM, QUESTION 7(b)

Between 5 am and 5 pm on 3 March 2009, the height, h, of the tide in a harbour was given by

$$h = 1 + 0.7\sin\frac{\pi}{6}t \qquad \text{for } 0 \le t \le 12,$$

where h is in metres and t is in hours, with $t = 0$ at 5 am.

a What is the period of the function h? 1 mark

b What was the value of h at low tide, and at what time did low tide occur? 2 marks

c A ship is able to enter the harbour only if the height of the tide is at least 1.35 m.

Find all times between 5 am and 5 pm on 3 March 2009 during which the ship was able to enter the harbour. 3 marks

TOPIC EXAM

Question 11 (3 marks)

Consider the function $y = \cos x$ in the domain $0 \le x \le 2\pi$, which is shown below.

On the same set of axes, sketch the graph of $y = 3\cos\left(2x + \frac{\pi}{3}\right)$. 3 marks

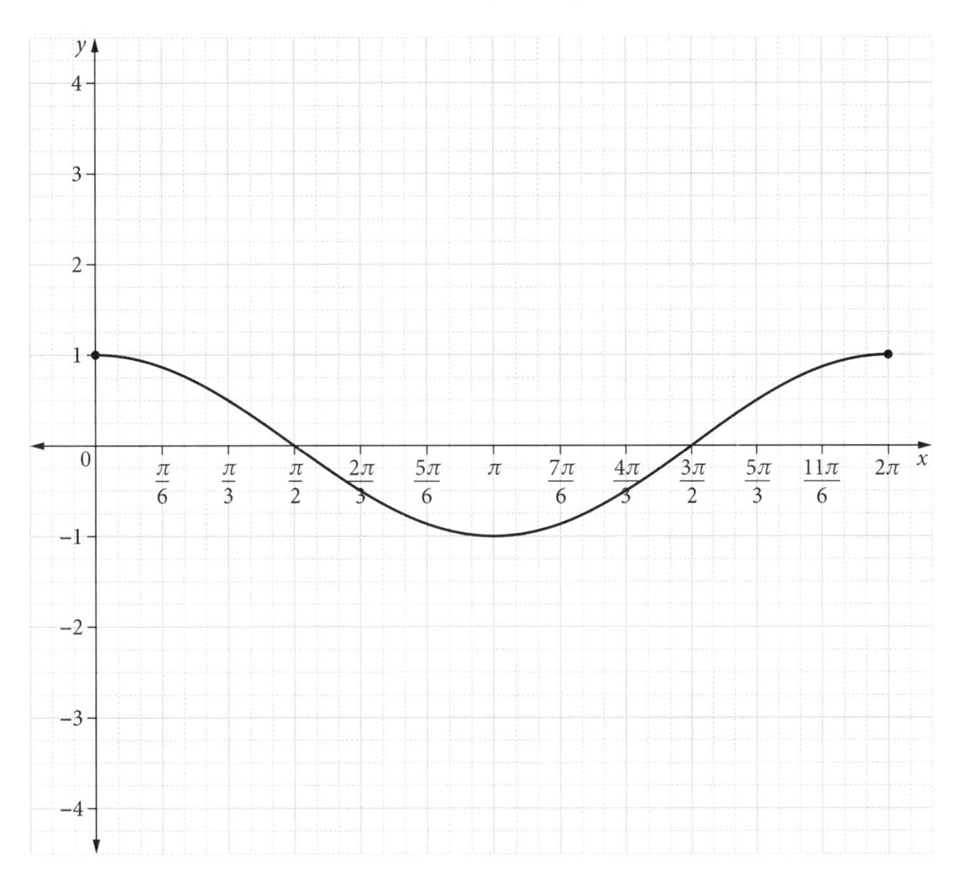

END OF PAPER

WORKED SOLUTIONS

Section I (1 mark each)

Question 1

A $\angle X = \frac{\pi}{2} - \frac{\pi}{6} = \frac{\pi}{3}$

$\tan\left(\frac{\pi}{3}\right) = \frac{x}{10}$

So $x = 10\tan\left(\frac{\pi}{3}\right)$.

Right-angled trigonometry. Make sure you are comfortable using radians as well as degrees.

Question 2

D If $\sin^2\theta = \frac{2}{9}$ then $\sin\theta = \pm\frac{\sqrt{2}}{3}$.

As $\sin\theta$ is both positive and negative, solutions will be found in all 4 quadrants.

Question 3

B $\sin\left(\frac{4\pi}{3}\right)$ and $\sin\left(\frac{\pi}{3}\right)$ are numerically equal but opposite in sign.

So $\sin\left(\frac{4\pi}{3}\right) = -\sin\left(\frac{\pi}{3}\right)$.

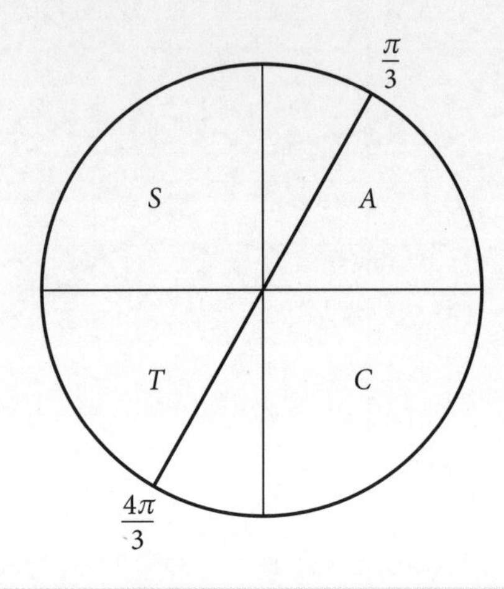

Remember 'All Stations To Central'.
Be comfortable using the unit circle. It is extremely useful.

Section II (✓ = 1 mark)

Question 4 (3 marks)

$\cos 30° = \frac{15}{AC}$

So $AC = \frac{15}{\cos 30°}$

$= \frac{15}{\frac{\sqrt{3}}{2}}$ ✓

$= \frac{30}{\sqrt{3}}$

$= \frac{30\sqrt{3}}{3}$

$= 10\sqrt{3}$ ✓

By Pythagoras' theorem,

$CD^2 = (10\sqrt{3})^2 + 10^2$

$= 400$

$CD = 20\text{ cm}$ ✓

Right-angled trigonometry concepts were studied in Years 9 and 10.

Question 5 (5 marks)

a $\angle AOB = 40 \times \frac{\pi}{180} = \frac{2\pi}{9}$ ✓

Straightforward conversion.

b $l_{AB} = r\theta$

$= 12 \times \frac{2\pi}{9}$ ✓

$= \frac{24\pi}{9}$

Perimeter $= 24 + \frac{8\pi}{3}$ cm ✓

Question asks for exact answer in terms of π so score 1 mark only if you rounded to 32.38.

c $A = \frac{1}{2}r^2\theta$

$= \frac{1}{2} \times 12^2 \times \frac{2\pi}{9}$ ✓

$= 16\pi\text{ cm}^2$ ✓

Parts **b** and **c** were solved using formulas from the reference sheet.

Question 6 (3 marks)

$$\frac{\sin\theta}{12} = \frac{\sin 50°}{10} \checkmark$$

$$\sin\theta = \frac{6\sin 50°}{5}$$
$$\approx 0.91925\ldots$$

So $\theta \approx 67°$ ✓ OR $180° - 67° = 113°$. ✓

Common content with Maths Standard 2. Remember the ambiguous case (2 answers).

Question 7 (1 mark)

If $\cos\theta = \sin(2\theta + 15°)$, then the angles are complementary.

$$\theta + (2\theta + 15) = 90$$
$$3\theta + 15 = 90$$
$$3\theta = 75$$
$$\theta = 25° \checkmark$$

Simple, but complementary angle results are often forgotten.

Question 8 (4 marks)

a $\text{LHS} = \cot^2\theta\sin\theta + \sin\theta$

$= \sin\theta(\cot^2\theta + 1)$ ✓

$= \frac{1}{\text{cosec}\,\theta} \times \text{cosec}^2\theta \quad \left(\sin\theta = \frac{1}{\text{cosec}\,\theta}\right)$

$= \text{cosec}\,\theta$ ✓

$= \text{RHS} \qquad (\cot^2\theta + 1 = \text{cosec}^2\theta)$

Many students find this hard. Practise in learning and using trigonometric identities is key. Not all of them are on the HSC exam reference sheet.

b $2\cot^2\theta\sin\theta + 2\sin\theta = -4$

$2(\cot^2\theta\sin\theta + \sin\theta) = -4$

$\cot^2\theta\sin\theta + \sin\theta = -2$

$\text{cosec}\,\theta = -2$, from the result in part **a**. ✓

$\sin\theta = -\frac{1}{2}$

So $\theta = -\frac{\pi}{6}$ or $-\frac{5\pi}{6}$. ✓

Question 9 (5 marks)

a $\angle BAC = 270° - 210°$ ✓

$= 60°$

270° – 210° is an 'explanation'. Don't write an essay for a 1-mark question.

b In ΔBAD, $\tan 30° = \frac{BD}{18}$

$\frac{1}{\sqrt{3}} = \frac{BD}{18}$

$BD = \frac{18}{\sqrt{3}}$

$= 6\sqrt{3}$ m ✓

Straightforward question.

c $BC^2 = 18^2 + 12^2 - 2(18)(12)\cos 60°$ ✓

$= 252$

So $BC = \sqrt{252}$

$= 6\sqrt{7}$ m ✓

Straightforward application of cosine rule in 3 dimensions. Note the link to part **a**.

d $\tan\theta = \frac{BD}{BC}$

$= \frac{6\sqrt{3}}{6\sqrt{7}}$

$= \frac{\sqrt{3}}{\sqrt{7}}$

So $\theta \approx 33°$. ✓

Note the link to parts **b** and **c**.

Question 10 (6 marks)

a Period $= \frac{2\pi}{\left(\frac{6}{\pi}\right)}$ (use of formula $T = \frac{2\pi}{n}$)

$= 2\pi \times \frac{6}{\pi}$

$= 12$ ✓

Learn the period formula. It is not on the HSC exam reference sheet.

b Low tide occurs when $\sin\left(\frac{\pi}{6}t\right) = -1$

$\frac{\pi}{6}t = \frac{3\pi}{2}$

$\frac{t}{6} = \frac{3}{2}$

So $t = 9$.

So low tide occurs 9 hours after 5 am, which is 2 pm. ✓

So $h = 1 + 0.7(-1)$

$= 0.3$ m ✓

Remember that –1 is the minimum value of the sine function. You do not need to use calculus to find the minimum here.

c

$$h = 1 + 0.7\sin\left(\frac{\pi}{6}t\right)$$

$$1.35 = 1 + 0.7\sin\left(\frac{\pi}{6}t\right) \checkmark$$

$$0.35 = 0.7\sin\left(\frac{\pi}{6}t\right)$$

$$\sin\left(\frac{\pi}{6}t\right) = \frac{1}{2}$$

$$\frac{\pi}{6}t = \frac{\pi}{6} \text{ or } \frac{5\pi}{6}$$

$$t = 1 \text{ or } 5 \checkmark$$

The ship can safely enter the harbour between 6 am and 10 am. ✓

In this 2009 HSC question, some students forgot to find the second value of t. Remember to use your values of t to determine the times of day. You must answer the question!

Question 11 (3 marks)

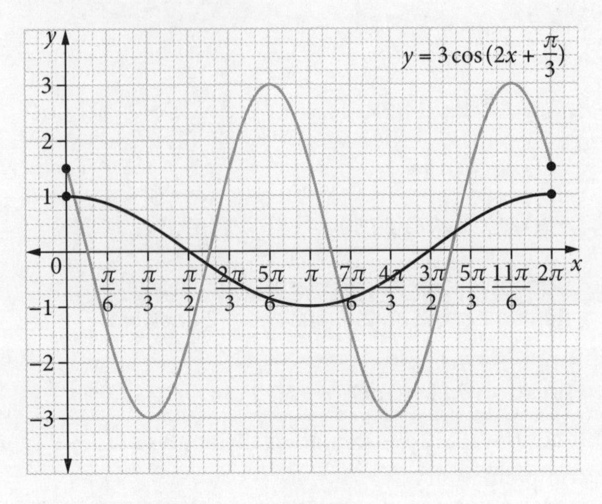

Apply the following sequence

1. $y = \cos x$
2. $y = 3\cos x$ (a vertical dilation of 3)
3. $y = 3\cos 2x$ (a horizontal dilation of $\frac{1}{2}$)
4. $y = 3\cos 2\left(x + \frac{\pi}{6}\right)$ (a horizontal translation of $\frac{\pi}{6}$ units to the left)

✓ if your graph has an amplitude of 3
✓ if your graph has a period of π
✓ if your graph begins and ends at $(0, 1.5)$ and $(2\pi, 1.5)$

HSC EXAM TOPIC GRID (2011–2020)

This table shows the coverage of this topic in past HSC exams by question number. The past exams can be downloaded from the NESA website (www.educationstandards.nsw.edu.au) by selecting 'Year 11 – Year 12', 'HSC exam papers'. NESA marking feedback and guidelines can also be found there.

Before 2020, 'Mathematics Advanced' was called 'Mathematics'. For these exams, select 'Year 11 – Year 12', 'Resources archive', 'HSC exam papers archive'.

	Transformations of trigonometric functions	Trigonometric equations	Applications of trigonometric functions
2011		2(b)	
2012		6	
2013	6, 13(a)	13(a)	13(a)
2014		7, 15(a)	
2015		12(a)	15(c)
2016	6	8, 11(g)	
2017	14(a)		
2018		15(a)(iii)	15(a)
2019	7		
2020 new course	6, 31		31

CHAPTER 3 TOPIC EXAM

3

Differentiation

MA-C2 Differential calculus

C2.1 Differentiation of trigonometric, exponential and logarithmic functions

C2.2 Rules of differentiation

MA-C3 Applications of differentiation

C3.1 The first and second derivatives

C3.2 Applications of the derivative

- A reference sheet is provided on page 195 at the back of this book
- For questions in Section II, show relevant mathematical reasoning and/or calculations

Reading time: 4 minutes
Working time: 1 hour
Total marks: 33

Section I – 3 questions, 3 marks
- Attempt Questions 1–3
- Allow about 5 minutes for this section

Section II – 6 questions, 30 marks
- Attempt Questions 4–9
- Allow about 55 minutes for this section

Section I

- Attempt Questions 1–3 **3 marks**
- Allow about 5 minutes for this section

Question 1

The graph of $y = f(x)$ is shown.

Which pair of inequalities accurately describes the sign of $f(x)$ and $f'(x)$ at the point P?

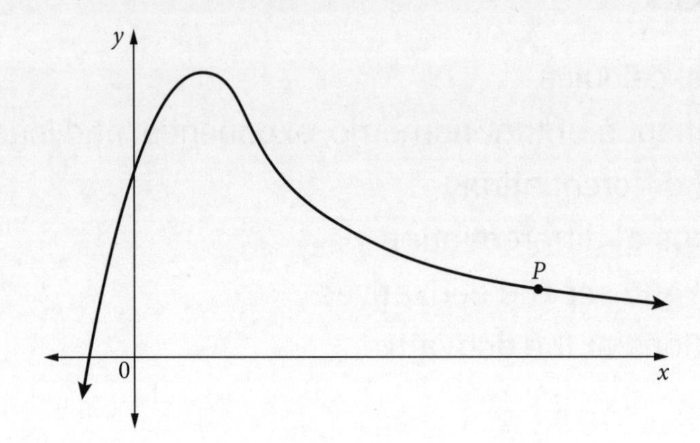

A $f(x) > 0$ and $f'(x) > 0$

B $f(x) > 0$ and $f'(x) < 0$

C $f(x) < 0$ and $f'(x) > 0$

D $f(x) < 0$ and $f'(x) < 0$

Question 2

Given that $g(x) = (3 - 2x)^8$, what is $g'(x)$?

A $g'(x) = 8(3 - 2x)^7$

B $g'(x) = -16(3 - 2x)^8$

C $g'(x) = 16(3 - 2x)^7$

D $g'(x) = -16(3 - 2x)^7$

Question 3

Find the gradient of the tangent drawn to the graph of $y = x^2 + 5x$ at the point $(-1, -4)$.

A -6

B -4

C 3

D 7

Section II

- Attempt Questions 4–9 **30 marks**
- Allow about 55 minutes for this section
- Answer the questions in the spaces provided. These spaces provide guidance for the expected length of response.
- Your responses should include relevant mathematical reasoning and/or calculations.

Question 4 (5 marks)

Differentiate each function.

a $y = \frac{2}{x}$ 1 mark

b $y = x^2 e^{2x}$ 2 marks

c $y = \frac{\ln x}{x}$ 2 marks

Question 5 (3 marks)

Find the equation of the tangent to the curve $y = \sqrt{1 - x}$ at the point where $x = -3$. 3 marks

Question 6 (2 marks)

Differentiate $f(x) = x^2 - 3x + 1$ from first principles. 2 marks

9780170459235

Question 7 (6 marks)

A particle is moving in a straight line. Its displacement, x m, from a fixed point, O, after t seconds is given by

$$x = 2t^2 - 5t,$$

where $t \geq 0$.

a Find an expression for the velocity in terms of t. 1 mark

b Find when the particle is at rest. 1 mark

c Show that the particle is moving with constant acceleration. 1 mark

d Find the velocity of the particle after 1 second. 1 mark

e By considering your answers to parts **c** and **d**, or otherwise, describe the motion of the particle at $t = 1$. 2 marks

Questions 4–7 are worth 16 marks in total (Section II halfway point)

Question 8 (8 marks)

A function is defined by $f(x) = \frac{1}{3}x^3 - x^2 - 8x$.

a Find the stationary points and determine their nature. 4 marks

b Show that there is a point of inflection at $\left(1,-8\frac{2}{3}\right)$. 2 marks

c Hence, sketch the graph of $f(x)$. 2 marks

Question 9 (6 marks)

The diagram shows the graph of $f(x) = \sqrt{12 - x}$.

A point $P(p, f(p))$ is chosen such that $0 < p < 12$.

Perpendiculars are drawn from P to both axes to form a rectangle, which has been shaded below.

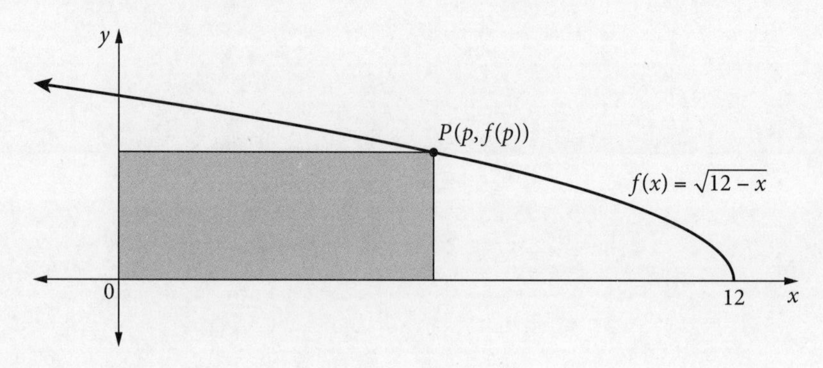

a Show that the area of the rectangle is given by $A = p\sqrt{12 - p}$. 1 mark

b Show that $\dfrac{dA}{dp} = \dfrac{24 - 3p}{2\sqrt{12 - p}}$. 2 marks

c Find the maximum possible area of the shaded rectangle. 3 marks

END OF PAPER

WORKED SOLUTIONS

Section I (1 mark each)

Question 1

B Point P is above the x-axis, so $f(x) > 0$.

The function is decreasing at point P, so $f'(x) < 0$.

Know the geometrical significance of both first and second derivatives.

Question 2

D Since $g(x)$ is in the form $[f(x)]^n$,

$$g'(x) = nf'(x)[f(x)]^{n-1}$$

That is, if $g(x) = (3 - 2x)^8$,

$$\begin{aligned} g'(x) &= 8(-2x)(3 - 2x)^7 \\ &= -16x(3 - 2x)^7 \end{aligned}$$

This is a direct application of the chain rule, which appears on the HSC exam reference sheet.

Question 3

C If $y = x^2 + 5x$, $\frac{dy}{dx} = 2x + 5$.

When $x = -1$,

$$\begin{aligned} \frac{dy}{dx} &= 2(-1) + 5 \\ &= 3 \end{aligned}$$

Common straightforward question.

Section II (✓ = 1 mark)

Question 4 (5 marks)

a If $y = \frac{2}{x} = 2x^{-1}$

$$\begin{aligned} y' &= -2x^{-2} \checkmark \\ &= -\frac{2}{x^2} \end{aligned}$$

Common question.

b $y = x^2e^{2x}$

$$\begin{aligned} \text{So } y' &= x^2(2e^{2x}) + e^{2x}(2x) \checkmark \\ &= 2xe^{2x}(x + 1) \checkmark \end{aligned}$$

Straightforward application of the product rule.

c $y = \frac{\ln x}{x}$

$$\begin{aligned} \text{So } y' &= \frac{x\left(\frac{1}{x}\right) - \ln x(1)}{x^2} \checkmark \\ &= \frac{1 - \ln x}{x^2} \checkmark \end{aligned}$$

Straightforward application of the quotient rule.

Question 5 (3 marks)

$$\begin{aligned} y &= \sqrt{1 - x} \\ &= (1 - x)^{\frac{1}{2}} \end{aligned}$$

$$\begin{aligned} \text{So } \frac{dy}{dx} &= \frac{1}{2}(1 - x)^{-\frac{1}{2}}(-1) \checkmark \\ &= \frac{-1}{2(1 - x)^{\frac{1}{2}}} \\ &= \frac{-1}{2\sqrt{1 - x}} \end{aligned}$$

When $x = -3$

$$\begin{aligned} y &= \sqrt{1 - (-3)} \\ &= 2 \end{aligned}$$

$$\begin{aligned} \frac{dy}{dx} &= \frac{-1}{2\sqrt{1 - (-3)}} \checkmark \\ &= -\frac{1}{4} \end{aligned}$$

$$\begin{aligned} y - y_1 &= m(x - x_1) \\ y - 2 &= -\frac{1}{4}(x + 3) \checkmark \end{aligned}$$

$$4y - 8 = -x - 3$$

So $x + 4y - 5 = 0$.

Although this is fairly straightforward, the function being a square root makes the differentiation harder. Still, students studying at the Maths Advanced level are expected to be able to deal confidently with questions of this standard. Working with fractional and negative indices is very important.

Question 6 (2 marks)

$f(x) = x^2 - 3x + 1$

So $f(x+h) = (x+h)^2 - 3(x+h) + 1$
$= x^2 + 2xh + h^2 - 3x - 3h + 1$ ✓

By first principles,

$$\begin{aligned} f'(x) &= \lim_{h\to 0}\frac{f(x+h)-f(x)}{h} \\ &= \lim_{h\to 0}\frac{x^2+2xh+h^2-3x-3h+1-(x^2-3x+1)}{h} \\ &= \lim_{h\to 0}\frac{2xh+h^2-3h}{h} \\ &= \lim_{h\to 0} 2x+h-3 \\ &= 2x-3 \ ✓ \end{aligned}$$

The formula for differentiation by first principles does not appear on the HSC exam reference sheet so must be memorised. Check your answer by the direct method.

Question 7 (6 marks)

a $v = \frac{dx}{dt} = 4t - 5$ ✓

Straightforward question.

b $0 = 4t - 5$
$4t = 5$
$t = \frac{5}{4}$ s ✓

The phrase 'at rest' means the particle is not moving, so $v = 0$.

c $a = \frac{dv}{dt} = 4\,\text{m/s}^2$ ✓

So acceleration is constant.

Straightforward question.

d $v = 4t - 5$

When $t = 1$:

$v = 4(1) - 5$
$= -1$ m/s ✓

Straightforward question.

e Since $v < 0$, we know that the particle is moving left. ✓

Since $a > 0$, we know the particle is slowing down because v and a have opposing signs. ✓

To determine if a particle is speeding up or slowing down, both the signs of v and a must be considered.

Question 8 (8 marks)

a $f(x) = \frac{1}{3}x^3 - x^2 - 8x$

So $f'(x) = x^2 - 2x - 8$. ✓

Stationary points occur where $f'(x) = 0$.

$0 = x^2 - 2x - 8$
$= (x-4)(x+2)$

So $x = -2$ and 4. ✓

$f(-2) = \frac{1}{3}(-2)^3 - (-2)^2 - 8(-2)$
$= 9\frac{1}{3}$

$f(4) = \frac{1}{3}(4)^3 - (4)^2 - 8(4)$
$= -26\frac{2}{3}$

Use $f''(x)$ to determine nature.

$f''(x) = 2x - 2$
So $f''(-2) = 2(-2) - 2$
$= -6$
< 0

So a local maximum occurs at $\left(-2, 9\frac{1}{3}\right)$. ✓

$f''(4) = 2(4) - 2$
$= 6$
> 0

So a local minimum occurs at $\left(4, -26\frac{2}{3}\right)$. ✓

Straightforward 4-mark curve-sketching question if you know the routine. Set your work out clearly. Show your substitutions and be careful working with negatives, particularly when they are being squared or cubed. Make it clear for the exam marker. Tell them what you are doing.

b $f''(x) = 2x - 2$

Point of inflection when $f''(x) = 0$.

$$0 = 2x - 2$$
$$2x = 2$$
$$x = 1$$

$$f(1) = \frac{1}{3}(1)^3 - (1)^2 - 8(1)$$
$$= \frac{1}{3} - 1 - 8$$
$$= -8\frac{2}{3}$$

There is a *possible* point of inflection at $\left(1, -8\frac{2}{3}\right)$. ✓

x	0	1	2
$f''(x)$	−2	0	2

So concavity changes at $\left(1, -8\frac{2}{3}\right)$ and, as such, it is confirmed as a point of inflection. ✓

When finding points of inflection, students often lose marks for not following all the correct steps. This is a 'show that' question, so more working must be shown as the answer is given. Note the detail shown in the evaluation of $f(1)$. The point of inflection is not 100% confirmed until it is established that concavity changes at that point. (A classic counterexample is $y = x^4$ at $x = 0$). Always put *values* in your table, never just '+' or '–'.

c

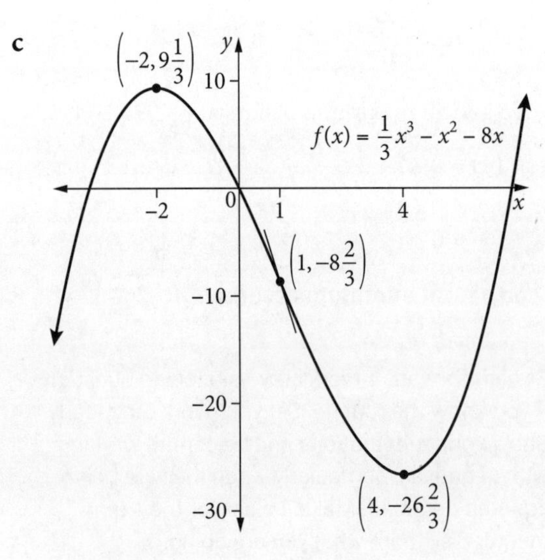

✓ for shape

✓ for showing the stationary points and point of inflection

Even though this is a 'sketch' and does not need to be precise, always rule your axes.

- The positions of your turning points should be roughly to scale.
- Label the key points.
- Try to draw a smooth curve.
- The y-intercept is usually easy to calculate. Show it.

As mentioned frequently by the HSC examiners, the quality of the graphs vary greatly and the best students draw graphs that are half a page and show clear details.

Question 9 (6 marks)

a $f(x) = \sqrt{12 - x}$

So $f(p) = \sqrt{12 - p}$.

$$\text{Area} = \text{base} \times \text{height}$$
$$= x \times f(x)$$
$$= pf(p)$$

So $A = p\sqrt{12 - p}$, as required. ✓

Optimisation questions are often aimed at Band 5/6 students. Again, 'show that' means extra detail is required. Convince the marker you could have achieved the result even if it were not given.

b $A = p\sqrt{12 - p}$

$$= p(12 - p)^{\frac{1}{2}}$$

So $\dfrac{dA}{dp} = p \times \dfrac{1}{2}(12 - p)^{-\frac{1}{2}} \times -1 + (12 - p)^{\frac{1}{2}} \times 1$ ✓

$$= \frac{-p}{2(12 - p)^{\frac{1}{2}}} + (12 - p)^{\frac{1}{2}}$$
$$= \frac{-p}{2\sqrt{12 - p}} + \sqrt{12 - p}$$
$$= \frac{-p}{2\sqrt{12 - p}} + \frac{2(12 - p)}{2\sqrt{12 - p}}$$
$$= \frac{-p + 2(12 - p)}{2\sqrt{12 - p}}$$
$$= \frac{-p + 24 - 2p}{2\sqrt{12 - p}}$$
$$= \frac{24 - 3p}{2\sqrt{12 - p}}$$, as required. ✓ (with process)

This is a heavy question in terms of the algebraic skills required and would be considered a Band 6 level question. Note that *every* part of the product rule is shown, including the value of u', which is 1. Also, note that no steps are skipped in the establishment of the result.

c A is maximised when $\frac{dA}{dp} = 0$.

$0 = \frac{24 - 3p}{2\sqrt{12 - p}}$

$0 = 24 - 3p$

So $p = 8$. ✓

p	7	8	9
$\frac{dA}{dp}$	$\frac{3}{2\sqrt{5}}$	0	$\frac{-3}{2\sqrt{3}}$

So A is maximised when $p = 8$. ✓

$A = p\sqrt{12 - p}$

$= 8\sqrt{12 - 8}$

$= 8 \times 2$

$A = 16 \text{ units}^2$ ✓

Even though the question asks for the maximum value of A, you must still show that the value of p that makes $\frac{dA}{dp} = 0$ *maximises* A and does not minimise it. The second derivative would be very messy to find, so revert to using a table of values to examine the value of $\frac{dA}{dp}$ on either side of $p = 8$. Again, write *values* in the table. The arrows shown are optional and are just there to give an impression of the gradients.

HSC exam topic grid (2011–2020)

This table shows the coverage of this topic in past HSC exams by question number. The past exams can be downloaded from the NESA (website www.educationstandards.nsw.edu.au) by selecting 'Year 11 – Year 12', 'HSC exam papers'. NESA marking feedback and guidelines can also be found there.

Before 2020, 'Mathematics Advanced' was called 'Mathematics'. For these exams, select 'Year 11 – Year 12', 'Resources archive', 'HSC exam papers archive'.

	Differentiation rules, tangents and normals	**Trigonometric, exponential and logarithmic functions**	**Stationary points, concavity and curve sketching**	**Optimisation* and motion problems**
2011	2(c)–(d), 4(a), 4(d)(i)	1(d), 2(d), 4(a)	7(a), 9(c)	7(b), 9(b)(i), 10(b)*
2012	11(c)–(d), 12(a)	11(d), 12(a), 14(c)	4, 14(a)	15(b)(ii), 16(b)*
2013	4, 11(c)–(d)	4, 11(c)–(d), 16(b)	8, 12(a)	10, 14(b)*
2014	11(c)	13(a)(i), 13(b), 14(a), 15(c)	9, 14(a), 14(e)	9, 13(c), 16(c)*
2015	11(e)–(f), 12(c), 12(e)(i)	6, 11(e)–(f), 15(a), 15(c)(i), 15(c)(iii)	13(c)	15(c)(iii)*, 16(c)*
2016	5, 11(b), 11(f), 12(d)(i), 16(b)	5, 11(f), 12(d)(i), 16(b)	13(a)	14(c)*, 16(a)(i)–(iii), 16(b)*
2017	3, 11(c)–(d), 12(a)	3, 11(c)–(d), 14(c)(i)	4, 9, 13(b)	10, 16(a)*
2018	5, 11(f)–(g), 12(b)	5, 11(f)–(g), 12(b)	9, 13, 14(c)	12(d), 16(a)*
2019	11(b)–(c), 13(c)(i), 16(c)(i)	11(b), 12(c)(ii), 13(c)(i)	8, 14(b)(i), 14(b)(iv)	8, 10, 15(c)*
2020 new course	10, 18(a), 21(b), 29	18(a), 21(b), 29, 31(c)	8, 16	25*

CHAPTER 4 TOPIC EXAM

4

Integration

MA-C4 Integral calculus

C4.1 The anti-derivative

C4.2 Areas and the definite integral

- A reference sheet is provided on page 195 at the back of this book
- For questions in Section II, show relevant mathematical reasoning and/or calculations

Reading time: 4 minutes
Working time: 1 hour
Total marks: 33

Section I – 3 questions, 3 marks
- Attempt Questions 1–3
- Allow about 5 minutes for this section

Section II – 10 questions, 30 marks
- Attempt Questions 4–13
- Allow about 55 minutes for this section

Section I

- Attempt Questions 1–3 **3 marks**
- Allow about 5 minutes for this section

Question 1

$f(x) = \sin x$ is an odd function.

Given $\int_0^2 \sin x \, dx \approx 1.42$, what is the value of $\int_{-2}^2 \sin x \, dx$?

A −2.84

B −1.42

C 0

D 2.84

Question 2

What is the value of $\int_{-1}^1 \sqrt{1 - x^2} \, dx$?

A 0

B $\frac{\pi}{2}$

D 2

C π

Question 3 ©NESA 2016 HSC EXAM, QUESTION 9

What is the value of $\int_{-3}^2 |x + 1| \, dx$?

A $\frac{5}{2}$

B $\frac{11}{2}$

C $\frac{13}{2}$

D $\frac{17}{2}$

9780170459235

Section II

- Attempt Questions 4–13 **30 marks**
- Allow about 55 minutes for this section
- Answer the questions in the spaces provided. These spaces provide guidance for the expected length of response.
- Your responses should include relevant mathematical reasoning and/or calculations.

Question 4 (2 marks)

a Find $\int (x^2 - x - 1)\,dx$. 1 mark

b Find $\int (3x + 1)^5\,dx$. 1 mark

Question 5 (3 marks)

Evaluate $\int_0^1 \frac{1}{(x+1)^2}\,dx$. 3 marks

9780170459235

Question 6 (2 marks) ©NESA 2011 HSC EXAM, QUESTION 4(c)

The gradient of a curve is $\frac{dy}{dx} = 6x - 2$. The curve passes through the point $(-1, 4)$.

What is the equation of the curve? 2 marks

Question 7 (2 marks)

Evaluate $\int_0^{\frac{\pi}{2}} \sin\left(\frac{x}{3}\right) dx$. 2 marks

Question 8 (2 marks)

Find $\int \frac{10x}{\sqrt{1-x^2}}\,dx$. 2 marks

Question 9 (2 marks)

Find the value of k if $\int_2^k \frac{2}{x-1}\,dx = \log_e 3$ and $k > 2$. 2 marks

Question 10 (4 marks) ©NESA 2014 HSC EXAM, QUESTION 12(d)

The parabola $y = -2x^2 + 8x$ and the line $y = 2x$ intersect at the origin and at the point A.

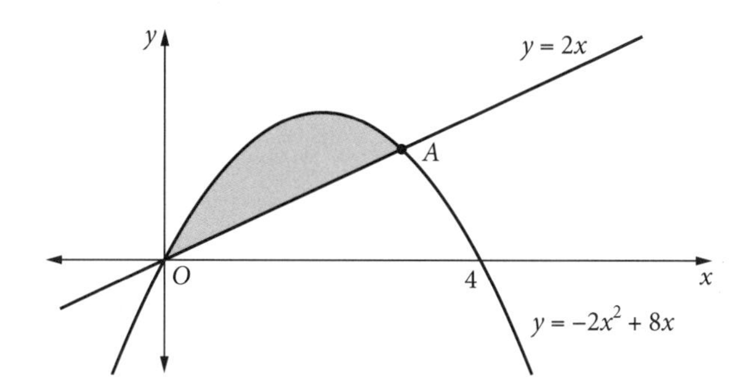

a Find the x-coordinate of the point A. 1 mark

b Calculate the area enclosed by the parabola and the line. 3 marks

Questions 4–10 are worth 17 marks in total (Section II halfway point)

Question 11 (4 marks)

The diagram below shows part of the graph of the function $f(x) = \frac{e^x}{x^3}$.

The area bounded by the graph and the x-axis between $x = 2$ and $x = 8$ has been shaded.

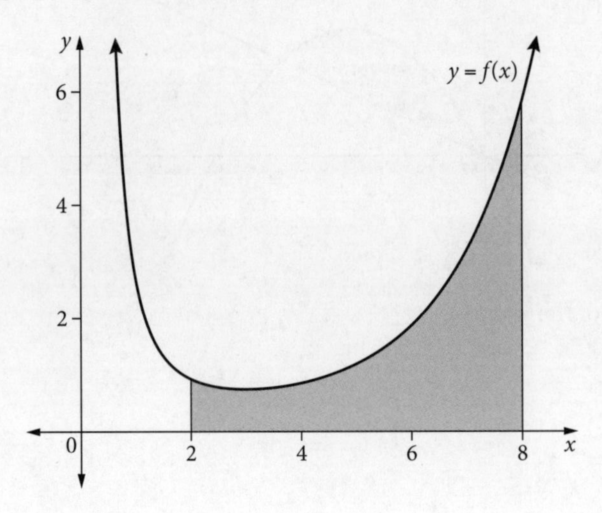

a Complete the table below. 1 mark

x	2	4	6	8
$f(x)$	0.92	0.85		5.82

b Using three applications of the trapezoidal rule, estimate the area of the shaded region, correct to one decimal place. 2 marks

c By considering the graph, or otherwise, determine with an appropriate justification, whether your answer in part **b** is greater than or less than the exact area. 1 mark

Question 12 (4 marks)

A particle is moving in a straight line. Its acceleration is given by $\ddot{x} = 6t - 3$.

Initially, the particle is stationary at the origin.

a Find an expression for $\dot{x}$, its velocity, in terms of t. 1 mark

b Find when the particle next comes to rest. 1 mark

c Find the distance travelled by the particle in the first 3 seconds. 2 marks

Question 13 (5 marks)

The rate at which people are being admitted to a concert is given by $\frac{dN}{dt} = t(60 - t)$, where t is the time in minutes since the gates were opened. That is, the venue was initially empty.

a At what rate were people being admitted 10 minutes after the gates were opened? 1 mark

b Find an expression for the number of people in the venue at any time t. 2 marks

c Determine the number of people who attended the concert. 2 marks

END OF PAPER

WORKED SOLUTIONS

Section I (1 mark each)

Question 1

C As $f(x)$ is odd,

$$\int_{-a}^{a} f(x)\,dx = 0$$

Property of odd functions. 'Negative' and positive areas cancel each other out.

Question 2

B $\int_{-1}^{1} \sqrt{1-x^2}\,dx$

$= \frac{1}{2} \times \pi \times 1^2$

$= \frac{\pi}{2}$

This is a semicircle of radius 1 unit.

Use the formula $A = \pi r^2$.

This function is difficult to integrate algebraically, so look for shortcuts involving the shape of the graph, such as the equation of a semicircle. This applies to Questions 1 to 3. Remember the connection between the integral and the area under a curve.

Question 3

C Consider the graph.

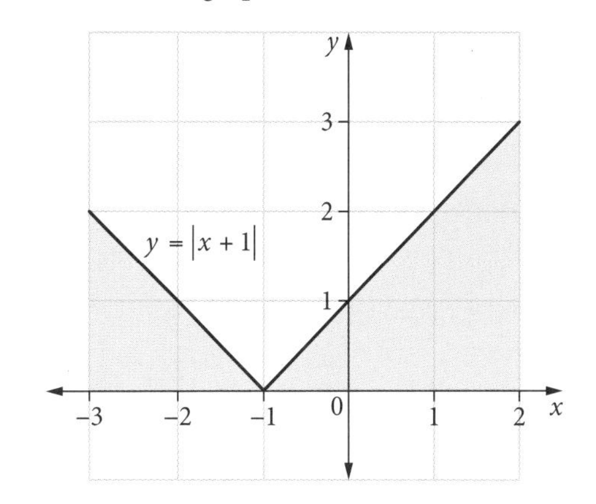

Find the shaded area to evaluate the integral. It is the sum of 2 triangles.

Area $= \frac{1}{2} \times 2 \times 2 + \frac{1}{2} \times 3 \times 3$

$= 6.5 \text{ units}^2$

So $\int_{-3}^{2} |x+1|\,dx = 6.5$

$= \frac{13}{2}$

Often in HSC exams, integration problems involving areas require measurement methods rather than formal integration.

Section II (✓ = 1 mark)

Question 4 (2 marks)

a $\int (x^2 - x - 1)\,dx$

$= \frac{1}{3}x^3 - \frac{1}{2}x^2 - x + c$ ✓

Straightforward integration.

b $\int (3x+1)^5\,dx$

$= \frac{(3x+1)^6}{6 \times 3} + c$ ✓

$= \frac{(3x+1)^6}{18} + c$

Application of the reverse chain rule.

Question 5 (3 marks)

$\int_0^1 \frac{1}{(x+1)^2}\,dx = \int_0^1 (x+1)^{-2}\,dx$ ✓

$= \left[\frac{(x+1)^{-1}}{-1}\right]_0^1$ ✓

$= \left[\frac{-1}{x+1}\right]_0^1$

$= -\frac{1}{2} - (-1)$ ✓

$= \frac{1}{2}$

This is not a 'log' integral. Good knowledge of negative indices is required, along with an ability to integrate using negative indices with the reverse chain rule. This is not an easy question but should be within the capability of a Mathematics Advanced student.

Question 6 (2 marks)

$\frac{dy}{dx} = 6x - 2$

So $y = 3x^2 - 2x + c$. ✓

As the curve passes through $(-1, 4)$, $f(-1) = 4$.

$4 = 3(-1)^2 - 2(-1) + c$
$4 = 3 + 2 + c$
$c = -1$ ✓

So $y = 3x^2 - 2x - 1$.

> Typical HSC question. Common student errors: forgetting '+ c', not finding the value of c, and substituting $(-1, 4)$ into $\frac{dy}{dx} = 6x - 2$.

Question 7 (2 marks)

$\int_0^{\frac{\pi}{2}} \sin\left(\frac{x}{3}\right) dx$

$= \left[-3\cos\left(\frac{x}{3}\right)\right]_0^{\frac{\pi}{2}}$ ✓

$= \left(-3\cos\frac{\pi}{6}\right) - (-3\cos 0)$

$= -\frac{3\sqrt{3}}{2} + 3$ ✓

$= 3 - \frac{3\sqrt{3}}{2}$

$= \frac{6 - 3\sqrt{3}}{2}$

> Definite integrals involving trigonometric functions are common in HSC exams. Be careful when substituting 0. Do not assume that the answer will be 0. Here, it is 3. This is a common mistake.

Question 8 (2 marks)

$\int \frac{10x}{\sqrt{1 - x^2}} dx$

$= \int \frac{10x}{(1 - x^2)^{\frac{1}{2}}} dx$ ✓

$= -5\int -2x(1 - x^2)^{-\frac{1}{2}} dx$

$= -5 \times \frac{1}{\frac{1}{2}}(1 - x^2)^{\frac{1}{2}} + c$ ✓

$= -10(1 - x^2)^{\frac{1}{2}} + c$

$= -10\sqrt{1 - x^2} + c$

> This question uses the reverse chain rule, which appears on the HSC exam reference sheet:
>
> $\int f'(x)[f(x)]^n dx = \frac{1}{n+1}[f(x)]^{n+1} + c$
>
> where $n \neq -1$.

Question 9 (2 marks)

If $\int_2^k \frac{2}{x - 1} dx = \log_e 3$ and $k > 2$, then

$\left[2\ln|x - 1|\right]_2^k = \ln 3$ ✓

$2\ln(k - 1) - 2\ln 1 = \ln 3$
$2\ln(k - 1) = \ln 3$
$\ln(k - 1)^2 = \ln 3$
$(k - 1)^2 = 3$
$k - 1 = \sqrt{3}$ (since $k > 2$)
$k = \sqrt{3} + 1$ ✓

> You must recognise the reciprocal function and know the result
>
> $\int \frac{f'(x)}{f(x)} dx = \ln|f(x)| + c$
>
> which appears on the HSC exam reference sheet.

Question 10 (4 marks)

a Solving simultaneously,

$-2x^2 + 8x = 2x$
$-2x^2 + 6x = 0$
$-2x(x - 3) = 0$
$x = 3$ $(x \neq 0)$ ✓

> Straightforward common HSC exam question on areas of integration.

b Area $\int_0^3 (-2x^2 + 8x - 2x) dx$ ✓

$= \int_0^3 (-2x^2 + 6x) dx$

$= \left[-\frac{2x^3}{3} + 3x^2\right]_0^3$ ✓

$= \left(-\frac{2(3^3)}{3} + 3(3^2)\right) - 0$

$= 9$ units^2 ✓

> 2014 HSC exam question. Common errors were not integrating properly or differentiating instead. Many similar questions in HSC exam papers. Note: Always show *both* substitutions in the definite integral, even if the second one is 0.

Question 11 (4 marks)

a

x	2	4	6	8
$f(x)$	0.92	0.85	1.87	5.82

✓

> Simple substitution question.

b Area $\approx \frac{h}{2}\left[f(2)+2\left(f(4)+f(6)\right)+f(8)\right]$

$= \frac{2}{2}\left[0.92+2(0.85+1.87)+5.82\right]$ ✓

≈ 12.8 ✓

The formula appears on the HSC exam reference sheet but without 'h'. Just remember, h is the width of a strip (interval). In this case, 2.

c Since the graph of $f(x)$ is concave up for $2 \le x \le 8$, the 3 trapeziums drawn over the curve will overestimate the exact area, so the area calculated in part **b** will be greater. ✓

Important to be able to identify this. Know how the trapezoidal rule works!

Question 12 (4 marks)

a $\dot{x} = \int \ddot{x}\, dt$

$= \int (6t-3)dt$

So $\dot{x} = 3t^2 - 3t + c$.

When $t = 0$, $\dot{x} = 0$, so $c = 0$.

So $\dot{x} = 3t^2 - 3t$. ✓

Typical question in integrating acceleration to find velocity. Note: Had this been a *'show that'* question, the substitution that led to the value of c would need to be shown.

b $\dot{x} = 3t^2 - 3t$

$0 = 3t^2 - 3t$

$0 = t^2 - t$

$0 = t(t-1)$

$t = 0, 1$

So the particle next comes to rest after 1 second. ✓

Solving an equation for zero velocity when the particle is 'at rest'.

c $x = \int \dot{x}\, dt$

$= \int (3t^2 - 3t)dt$

$= t^3 - \frac{3t^2}{2} + K$

When $t = 0$, $x = 0$, so $K = 0$.

$x = t^3 - \frac{3t^2}{2}$ ✓

From the initial conditions we know that $x = 0$ when $t = 0$.

When $t = 1$, $x = 1^3 - \frac{3\times 1^2}{2} = -\frac{1}{2}$.

When $t = 3$, $x = 3^3 - \frac{3\times 3^2}{2} = 13\frac{1}{2}$.

The particle begins at 0, travels left to $x = -\frac{1}{2}$ and then right to $13\frac{1}{2}$; a total of $14\frac{1}{2}$ metres. ✓

In these types of problems, the particle stops to turn around in the vast majority of cases, so we need to take this into account when calculating *distance* rather than displacement. We cannot just integrate velocity between $t = 0$ and 3.

So, in part **b**, we see the particle stops at $t = 1$.

Therefore, find the position at the particle at $t = 0$, 1 and 3.

The journey is shown roughly below.

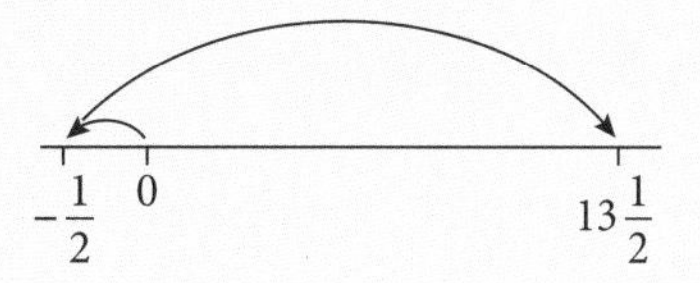

Question 13 (5 marks)

a When $t = 10$,

$\frac{dN}{dt} = 10(60-10)$

$= 500$ people per minute. ✓

Straightforward rates question.

b $\frac{dN}{dt} = t(60-t)$

$= 60t - t^2$

So $N = 30t^2 - \frac{t^3}{3} + c$. ✓

When $t = 0$, $N = 0$, so $c = 0$.

So $N = 30t^2 - \frac{t^3}{3}$. ✓

Typical integration question.

c The total number of people attending the concert will be the maximum value of N, or when the rate of change $N' = 0$.

$$\frac{dN}{dt} = t(60 - t) = 0$$

So $t = 0$ or 60.

That is, when $t = 60$, people are no longer entering the venue.

When $t = 60$, ✓

$$N = 30(60)^2 - \frac{60^3}{3}$$
$$= 36\,000$$ ✓

This is a complex question testing deep conceptual understanding and applications of differentiation. It is important to realise that the rate of change will be zero just before people start entering the venue and when they stop entering the venue.

HSC exam topic grid (2011–2020)

This table shows the coverage of this topic in past HSC exams by question number. The past exams can be downloaded from the NESA website (www.educationstandards.nsw.edu.au) by selecting 'Year 11 – Year 12', 'HSC exam papers'. NESA marking feedback and guidelines can also be found there.

Before 2020, 'Mathematics Advanced' was called 'Mathematics'. For these exams, select 'Year 11 – Year 12', 'Resources archive', 'HSC exam papers archive'.

	Integration	Trapezoidal rule	Area under curves	Applications of integration
2011	2(e), 4(b), 4(d), 6(c)(ii)	5(c)*	6(c)	**4(c)**
2012	9, 11(g), 12(b)	12(d)*	10, 13(b)	15(b)(iii)–(iv)#
2013	11(e)–(f)	15(a)(i)	13(b), 14(d)	14(a)#, 16(a)
2014	4, 11(d)–(e), 13(a)	16(a)*	**12(d)**	11(f)
2015	11(g)–(h)	5	7, 9, 10, 16(a)	9#, 14(a)#, 15(c)
2016	**9**, 11(d), 12(d)	14(a)*	**9**, 13(e)	16(a)(iv)#
2017	11(b), 14(b)(i)	14(b)(ii)	14(d)	9, 13(d), 15(c)#
2018	11(e)		7, 10, 15(b)–(c)(i)	
2019	9, 11(e), 13(c)	16(b)*	12(d), 16(c)	8, 14(a)#, 14(b)(ii), 16(b)#
2020 new course	4, 13, 17, 18	20	7, 30	20

Questions in **bold** can be found in this chapter.

* Uses Simpson's rule, which is no longer in the course, but the trapezoidal rule can be used instead (with a similar answer)

Motion

CHAPTER 5 TOPIC EXAM

Series, investments, loans and annuities

MA-M1 Modelling financial situations

- M1.1 Modelling investments and loans
- M1.2 Arithmetic sequences and series
- M1.3 Geometric sequences and series
- M1.4 Financial applications of sequences and series

- A reference sheet is provided on page 195 at the back of this book
- For questions in Section II, show relevant mathematical reasoning and/or calculations

Reading time: 4 minutes
Working time: 1 hour
Total marks: 33

Section I – 3 questions, 3 marks
- Attempt Questions 1–3
- Allow about 5 minutes for this section

Section II – 6 questions, 30 marks
- Attempt Questions 4–9
- Allow about 55 minutes for this section

Section I

- Attempt Questions 1–3 **3 marks**
- Allow about 5 minutes for this section

Question 1

What is the 12th term of the arithmetic sequence $-4, -1, 2, \ldots$?

A 26

B 29

C 32

D 37

Question 2

Which of the following best describes the sequence $5, 5, 5, \ldots$?

A Arithmetic

B Geometric

C Neither arithmetic nor geometric

D Both arithmetic and geometric

Question 3

\$5000 is invested into an account earning 3% per annum for 3 years. Interest is compounded quarterly.

Which expression correctly evaluates the amount of interest earned?

A 5000×1.0075^{12}

B $5000 \times 1.03 \times 3$

C $5000 \times 1.03^{3} - 5000$

D $5000 \times 1.0075^{12} - 5000$

9780170459235

Section II

- Attempt Questions 4–9 **30 marks**
- Allow about 55 minutes for this section
- Answer the questions in the spaces provided. These spaces provide guidance for the expected length of response.
- Your responses should include relevant mathematical reasoning and/or calculations.

Question 4 (3 marks)

The 11th term of an arithmetic sequence is 30.

If the sum of the first 8 terms is 266, find the common difference. 3 marks

Question 5 (2 marks)

The table shows the future value of an annuity of $1 for a selection of interest rates per period and investment terms. The contributions are made at the end of each period.

	Interest rate per period							
Period	**1%**	**2%**	**3%**	**4%**	**5%**	**6%**	**7%**	**8%**
1	1.0000	1.0000	1.0000	1.0000	1.0000	1.0000	1.0000	1.0000
2	2.0100	2.0200	2.0300	2.0400	2.0500	2.0600	2.0700	2.0800
3	3.0301	3.0604	3.0909	3.1216	3.1525	3.1836	3.2149	3.2464
4	4.0604	4.1216	4.1836	4.2465	4.3101	4.3746	4.4399	4.5061
5	5.1010	5.2040	5.3091	5.4163	5.5256	5.6371	5.7507	5.8666
6	6.1520	6.3081	6.4684	6.6330	6.8019	6.9753	7.1533	7.3359
7	7.2135	7.4343	7.6625	7.8983	8.1420	8.3938	8.6540	8.9228
8	8.2857	8.5830	8.8923	9.2142	9.5491	9.8975	10.2598	10.6366
9	9.3685	9.7546	10.1591	10.5828	11.0266	11.4913	11.9780	12.4876
10	10.4622	10.9497	11.4639	12.0061	12.5779	13.1808	13.8164	14.4866
11	11.5668	12.1687	12.8078	13.4864	14.2068	14.9716	15.7836	16.6455
12	12.6825	13.4121	14.1920	15.0258	15.9171	16.8699	17.8885	18.9771
13	13.8093	14.6803	15.6178	16.6268	17.7130	18.8821	20.1406	21.4953
14	14.9474	15.9739	17.0863	18.2919	19.5986	21.0151	22.5505	24.2149
15	16.0969	17.2934	18.5989	20.0236	21.5786	23.2760	25.1290	27.1521
16	17.2579	18.6393	20.1569	21.8245	23.6575	25.6725	27.8881	30.3243

a Andy invests $800 at the end of each quarter for 3 years in an account earning 8% p.a. with interest compounded quarterly.

Find the future value of his annuity. 1 mark

b Find the present value of Andy's annuity. 1 mark

Question 6 (5 marks)

Consider the infinite series $24 + 16 + \frac{32}{3} + \ldots$

a Find its common ratio. 1 mark

b Find its limiting sum. 1 mark

c Find the first term of the series whose value is less than $\frac{1}{2}$. 3 marks

Question 7 (5 marks)

Vin plants rows of lettuces on his farm.

The first row contains 5 lettuce plants and each subsequent row has 2 more plants than the previous row.

a How many lettuces will there be in the 8th row? 1 mark

b Find the total number of lettuces in the first 8 rows. 1 mark

c If Vin plants a total of 700 lettuces, how many complete rows will he have planted? 3 marks

Questions 4–7 are worth 15 marks in total (Section II halfway point)

Question 8 (10 marks)

At Eastpac Bank, the Super Saver Annuity account earns 9.6% p.a. with interest compounded monthly.

a Explain why this is equivalent to 0.8% per month. 1 mark

b Let A_n represent the balance in the account after the nth instalment of interest has been added.

Show that if $\$P$ is invested at the beginning of each month for n months, then 3 marks

$$A_n = 126P(1.008^n - 1).$$

c Alexis invests \$150 at the beginning of each month for 36 months.

Find the final balance in her account to the nearest cent. 1 mark

d Simran wants the final value of his investment to be \$9000 so he can buy a car.

If he also invests for 36 months, find the size of the monthly contribution required to achieve his goal. 2 marks

e Oliver makes \$150 monthly contributions like Alexis, but wishes to achieve the same \$9000 goal as Simran.

For how many months must Oliver invest to achieve his goal? 3 marks

Question 9 (5 marks) ©NESA 2013 HSC EXAM, QUESTION 13(d)

A family borrows \$500 000 to buy a house. The loan is to be repaid in equal monthly instalments. The interest, which is charged at 6% per annum, is reducible and calculated monthly. The amount owing after n months, $\$A_n$, is given by

$$A_n = Pr^n - M(1 + r + r^2 + \cdots + r^{n-1}), \qquad \text{(Do NOT prove this)}$$

where P is the amount borrowed, $r = 1.005$ and M is the monthly repayment.

a The loan is to be repaid over 30 years. Show that the monthly repayment is \$2998 to the nearest dollar. 2 marks

b Show that the balance owing after 20 years is \$270 000 to the nearest thousand dollars. 1 mark

c After 20 years, the family borrows an extra amount, so that the family then owes a total of \$370 000. The monthly repayment remains at \$2998, and the interest rate remains the same.

How long will it take to repay the \$370 000? 2 marks

END OF PAPER

WORKED SOLUTIONS

Section I (1 mark each)

Question 1

B $T_n = a + (n-1)d$
$= -4 + 11 \times 3$
$= 29$

Straightforward application of a formula that appears on the HSC exam reference sheet.

Question 2

D To be an arithmetic sequence, $T_3 - T_2 = T_2 - T_1$.

$5 - 5 = 5 - 5 = 0$, so the sequence is arithmetic.

To be a geometric sequence, $\frac{T_3}{T_2} = \frac{T_2}{T_1}$.

$\frac{5}{5} = \frac{5}{5} = 1$, so the sequence is geometric.

Know the conditions for both types of series.

Question 3

D First, and crucially, the interest is compounded *quarterly*.

So $r = 0.03 \div 4 = 0.0075$ and $n = 12$.

Use the future value (compound interest) formula,

$A = 5000(1 + 0.0075)^{12}$.

So interest $= 5000(1 + 0.0075)^{12} - 5000$.

Always scan the question for the frequency of compounding.

Section II (✓ = 1 mark)

Question 4 (3 marks)

$T_n = a + (n-1)d$

$T_{11} = a + 10d$

$30 = a + 10d$ [1] ✓

$S_n = \frac{n}{2}[2a + (n-1)d]$

$S_8 = \frac{8}{2}[2a + 7d]$

$266 = 4(2a + 7d)$
$133 = 2(2a + 7d)$ ✓
$133 = 4a + 14d$ [2]

Solving simultaneously,

From [1]:
$a = 30 - 10d$

Substitute into [2]:
$133 = 4(30 - 10d) + 14d$
$= 120 - 40d + 14d$
$13 = -26d$

So $d = -\frac{1}{2}$. ✓

Questions requiring the use of simultaneous equations are common in HSC exams.

See 2018 – Q11(d) and 2017 – Q12(c). This question requires the use of T_n and S_n formulas, both of which are on the reference sheet.

Question 5 (2 marks)

a Future value $= 800 \times 13.4121$
$\approx \$10\,729.68$ ✓

b $A = P(1 + r)^n$
$10\,729.68 = P(1 + 0.02)^{12}$
So $P = \$8460.28$ ✓

Both parts **a** and **b** are straightforward and common content with Maths Standard 2.

Question 6 (5 marks)

a $r = \frac{T_2}{T_1}$
$= \frac{16}{24}$
$= \frac{2}{3}$ ✓

Straightforward question.

b $S = \dfrac{a}{1-r}$

$= \dfrac{24}{1-\frac{2}{3}}$

$= 72$ ✓

Straightforward application of a formula on the HSC exam reference sheet.

c $T_n = ar^{n-1}$

So $T_n = 24\left(\dfrac{2}{3}\right)^{n-1}$

We require, $24\left(\dfrac{2}{3}\right)^{n-1} < \dfrac{1}{2}$ ✓

$$\left(\frac{2}{3}\right)^{n-1} < \frac{1}{48}$$

$$\ln\left(\frac{2}{3}\right)^{n-1} < \ln\left(\frac{1}{48}\right)$$

$$(n-1)\ln\left(\frac{2}{3}\right) < \ln\left(\frac{1}{48}\right)$$

$$n-1 > \frac{\ln\left(\frac{1}{48}\right)}{\ln\left(\frac{2}{3}\right)} \quad [*]$$

$$n-1 > 9.54\ldots$$

$$n > 10.54\ldots$$ ✓

So T_{11} is the first term whose value is less than $\dfrac{1}{2}$.

$T_{11} = 24\left(\dfrac{2}{3}\right)^{11-1}$

$= \dfrac{8192}{19\,683}$ ✓

$= 0.416\ldots$

This is a more complex question aimed at Band 6 students. You must be able to interpret the question and know you are required to solve $T_n < \dfrac{1}{2}$ and use the logarithm laws. Note that at [*], the direction of the inequality sign changes, as $\ln\left(\dfrac{2}{3}\right) < 0$. Solving $24\left(\dfrac{2}{3}\right)^{n-1} = \dfrac{1}{2}$ instead of $24\left(\dfrac{2}{3}\right)^{n-1} < \dfrac{1}{2}$ looks easier and is technically not correct, but then you have to decide what to do with $n = 10.54\ldots$ Also, make sure that you actually answer the question by evaluating the term required.

Question 7 (5 marks)

a $T_n = a + (n-1)d$

So $T_8 = 5 + (8-1) \times 2$ ✓

$= 19$ lettuces

b $S_n = \dfrac{n}{2}(a + l)$

$S_8 = \dfrac{8}{2}(5 + 19)$ ✓

$= 96$ lettuces

For parts **a** and **b**, straightforward application of a formula on the HSC exam reference sheet.

c We know that $S_n = 700$.
We are required to find n.

$S_n = \dfrac{n}{2}[2a + (n-1)d]$

$700 = \dfrac{n}{2}[10 + (n-1)2]$ ✓

$700 = \dfrac{n}{2}[2n + 8]$

$700 = n(n + 4)$

$= n^2 + 4n$

So $n^2 + 4n - 700 = 0$.

Apply the general quadratic formula as the quadratic does not factorise.

$$n = \frac{-b \pm \sqrt{b^2 - 4ac}}{2a}$$

$$= \frac{-4 \pm \sqrt{4^2 - (4 \times 1 \times -700)}}{2}$$

$$= \frac{-4 \pm \sqrt{2816}}{2}$$ ✓

But $n > 0$, so

$n = \dfrac{-4 + \sqrt{2816}}{2}$ only.

So $n \approx 24.53$.

Vin will have planted 24 complete rows of lettuces. ✓

This question tests the same skills tested in the 2012 HSC exam paper in Q12(c). Higher level skills targeting Band 5/6.
Award third mark only if you have specifically mentioned that $n > 0$ and so $n = \dfrac{-4 - \sqrt{2816}}{2}$ has no meaning in the context of the question.

Question 8 (10 marks)

a Because $9.6\% \div 12 = 0.8\%$ ✓

b $A_1 = P(1 + 0.008)$
$= P(1.008)$

$A_2 = [P(1.008) + P]1.008$
$= P(1.008^2) + P(1.008)$
$= P(1.008^2 + 1.008)$ ✓

$A_3 = [P(1.008^2) + P(1.008) + P]1.008$
$= P(1.008^3) + P(1.008^2) + P(1.008)$
$= P(1.008^3 + 1.008^2 + 1.008)$

Generalising this pattern, we get:

$$A_n = P(1.008 + 1.008^2 + 1.008^3 + \cdots + 1.008^n) \checkmark$$
$$= P\left[\frac{1.008(1.008^n - 1)}{1.008 - 1}\right]$$
$$= P\left[\frac{1.008(1.008^n - 1)}{0.008}\right]$$
$$= 126P(1.008^n - 1) \text{ as } \frac{1.008}{0.008} = 126.$$

This line demonstrates an understanding of where the '126' given in the question comes from. It helps the marker know that you were not relying on the result being provided to you.

So $A_n = 126P(1.008^n - 1)$, as required. ✓

Parts **a** and **b** are straightforward questions. Defining A_n is best practice. The question will often do this. 2 months are normally enough although 3 are better to establish the pattern.

c In part **b**, we established $126P(1.008^n - 1)$.

If $P = 150$ and $n = 36$,

$A_{36} = 126 \times 150 \times (1.008^{36} - 1)$
$\approx \$6279.14$ ✓

d $A_{36} = 9000$

$9000 = 126P(1.008^{36} - 1)$ ✓

$$P = \frac{9000}{126(1.008^{36} - 1)}$$
$\approx \$215.00$ ✓

Parts **c** and **d** are straightforward applications of the formula.

e $P = 150, A_n = 9000, n = ?$

$A_n = 126P(1.008^n - 1)$ ✓
$9000 = 126 \times 150 \times (1.008^n - 1)$
$\frac{10}{21} = 1.008^n - 1$
$1.008^n = \frac{31}{21}$ ✓

$$\text{So } n = \frac{\ln\left(\frac{31}{21}\right)}{\ln(1.008)}$$
$= 48.8775\ldots$
≈ 49 months ✓

Finding n is always slightly more difficult because the solution requires logarithms but this type of question has been asked often.

Question 9 (5 marks)

a $A_n = Pr^n - M(1 + r + r^2 + \cdots + r^{n-1})$

The loan is to be repaid in 360 months, so $A_{360} = 0$.

$$0 = 500\,000(1.005)^{360} - M(1 + 1.005 + 1.005^2 + \ldots + 1.005^{359}) \ ✓$$

$$0 = 500\,000(1.005)^{360} - M\left(\frac{1.005^{360} - 1}{1.005 - 1}\right)$$

$$M\left(\frac{1.005^{360} - 1}{0.005}\right) = 500\,000(1.005)^{360}$$

$$M = \frac{500\,000(1.005)^{360} \times 0.005}{1.005^{360} - 1}$$

$= 2997.7526\ldots$ by calculator

$\approx \$2998$, as required ✓ with unrounded answer shown

This question from the 2013 HSC exam targets students at a Band 5/6 standard of achievement. Note that the formula for A_n has been provided in this question, so it does not need to be proven by the student. However, in an exam, you may be asked to prove the formula. Most students scored full marks for part **a**, with the common error of using $n = 30$ (not 360) and not writing the unrounded answer in the working.

b $A_n = Pr^n - M(1 + r + r^2 + \cdots + r^{n-1})$

$$A_{240} = 500\,000(1.005)^{240} - 2998(1 + 1.005 + 1.005^2 + \ldots + 1.005^{239})$$

$$= 500\,000(1.005)^{240} - 2998\left(\frac{1.005^{240} - 1}{0.005}\right) \quad [*]$$

$= \$269\,903.6342\ldots$ by calculator

$\approx \$270\,000$, as required ✓ with unrounded answer shown

Straightforward question. Even if you were unable to establish that $M = \$2998$ in part **a**, you can still use the result to answer part **b**. Always show your unrounded answer in a 'show that' question. It proves to the marker that you actually calculated the expression in the line marked [*].

c $A_n = Pr^n - M(1 + r + r^2 + \cdots + r^{n-1})$

Assume that the \$370 000 is repaid in k months.

Then,

$$A_k = 370\,000(1.005)^k - 2998\left(\frac{1.005^k - 1}{0.005}\right) \ ✓$$

$$= 370\,000(1.005)^k - 599\,600(1.005^k - 1)$$

$$= 370\,000(1.005)^k - 599\,600(1.005)^k + 599\,600 = 0$$

$$-229\,600(1.005)^k = -599\,600$$

$(1.005)^k = 2.6114\ldots$ by calculator

So $k = \dfrac{\log(2.6114\ldots)}{\log(1.005)}$

$= 192.4643\ldots$ by calculator

So the loan will be repaid in 193 months. ✓

This part was more challenging for the 2013 HSC students. One mark was awarded for realising that $A_k = 0$. Again, finding n (or k in this case) is a very common question requiring precise algebraic manipulations and the use of logarithms. You need more practise to have a deep understanding of these types of problems involving financial applications of series.

HSC exam topic grid (2011–2020)

This table shows the coverage of this topic in past HSC exams by question number. The past exams can be downloaded from the NESA website (www.educationstandards.nsw.edu.au) by selecting 'Year 11 – Year 12', 'HSC exam papers'. NESA marking feedback and guidelines can also be found there.

Before 2020, 'Mathematics Advanced' was called 'Mathematics'. For these exams, select 'Year 11 – Year 12', 'Resources archive', 'HSC exam papers archive'.

	Arithmetic sequences and series	Geometric sequences and series	Loans and annuities by series	Investments, loans and annuities (introduced in 2020)	
2011	3(a)	5(a)	8(c)	22, 23(c)	General Maths exams
2012	12(c)	15(a)	15(c)	9, 24	
2013		12(c)	**13(d)**	9, 26(e)	
2014	12(a), 14(d)	8	16(b)	21, 30(a)	Maths General 2 exams
2015	3	11(d)	14(c)	17, 26(d), 29(b), 30(c)	
2016	14(e)	14(b), 14(d)		8, 27(d), 28(d)	
2017	12(c)	16(b)	15(b)	10, 27(c), 28(c)	
2018	11(d), 14(d)	14(d)	16(c)	19, 26(c), 29(e)	
2019	12(b)	11(d)	16(a)	3, 9, 13, 42	Maths Standard 2 exam
2020 new course	12			26	

Questions in **bold** can be found in this chapter.

CHAPTER 6 TOPIC EXAM

Statistics and bivariate data

MA-S2 Descriptive statistics and bivariate data analysis

S2.1 Data (grouped and ungrouped) and summary statistics

S2.2 Bivariate data analysis

MA-S3 Random variables

S3.2 The normal distribution

Note: This topic exam includes z-scores and the normal distribution from the Probability distributions topic.

- A reference sheet is provided on page 195 at the back of this book
- For questions in Section II, show relevant mathematical reasoning and/or calculations

Reading time: 4 minutes
Working time: 1 hour
Total marks: 33

Section I – 3 questions, 3 marks
- Attempt Questions 1–3
- Allow about 5 minutes for this section

Section II – 6 questions, 30 marks
- Attempt Questions 4–9
- Allow about 55 minutes for this section

Section I

- Attempt Questions 1–3 **3 marks**
- Allow about 5 minutes for this section

Question 1

A group of people were asked to count the number of sports they had played from a provided list.

Their responses are listed in ascending order below.

0 0 1 1 1 2 3 3 4 5 5 5 6 6 7 8

Find the median and range of this data set.

A Median 3 and interquartile range 4.5

B Median 3.5 and interquartile range 4.5

C Median 3.5 and interquartile range 8

D Median 3.5 and interquartile range 5

Question 2

A data set has a mean of 50 and a standard deviation of 7.

A value of 40 is added to the data set.

How does the new value affect the mean and standard deviation of the data set?

A The mean increases and the standard deviation increases.

B The mean increases and the standard deviation decreases.

C The mean decreases and the standard deviation increases.

D The mean decreases and the standard deviation decreases.

Question 3

Which scatterplot most accurately represents two variables with a correlation coefficient of –0.92?

A

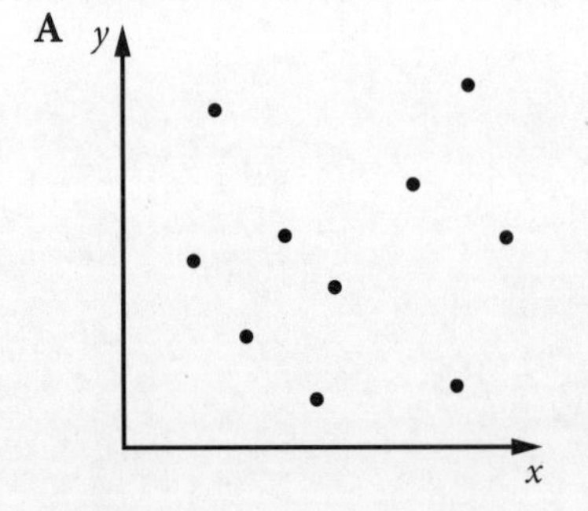

B

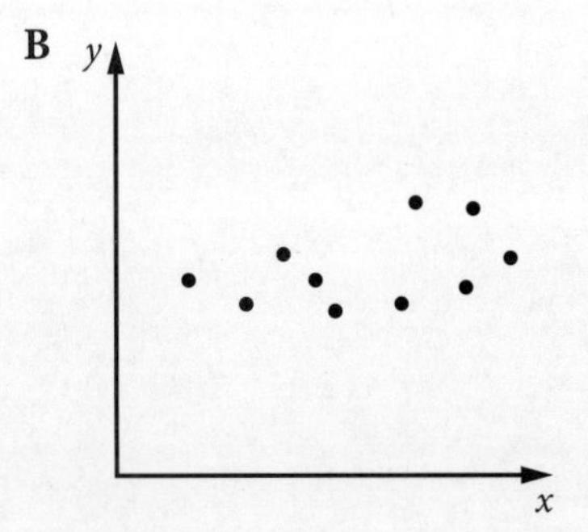

C

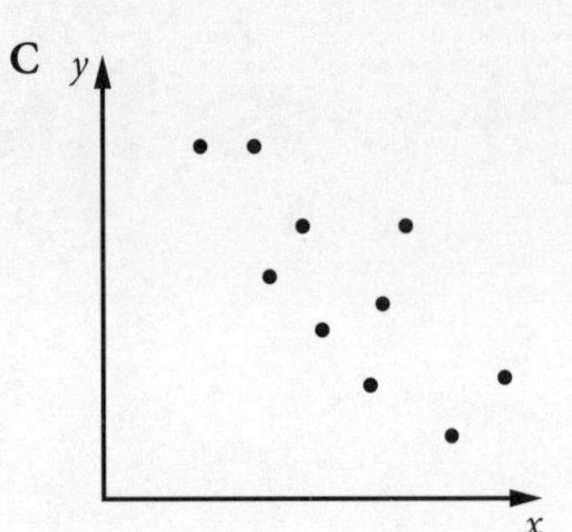

D

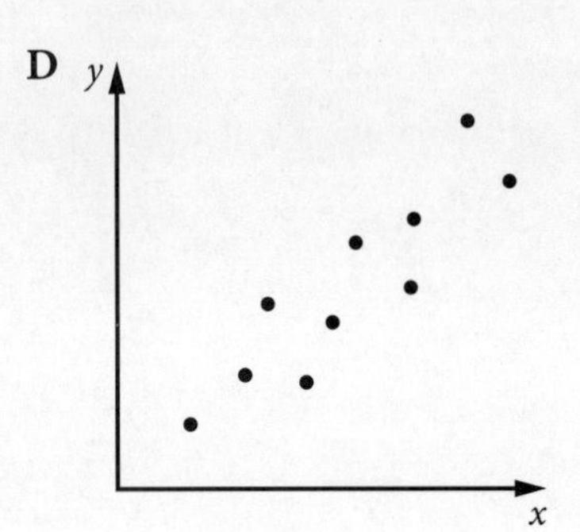

Section II

- Attempt Questions 4–9 **30 marks**
- Allow about 55 minutes for this section
- Answer the questions in the spaces provided. These spaces provide guidance for the expected length of response.
- Your responses should include relevant mathematical reasoning and/or calculations.

Question 4 (3 marks)

Frank owns a local pizza shop and records the types of pizzas he sells.

An incomplete Pareto chart displaying the data is shown.

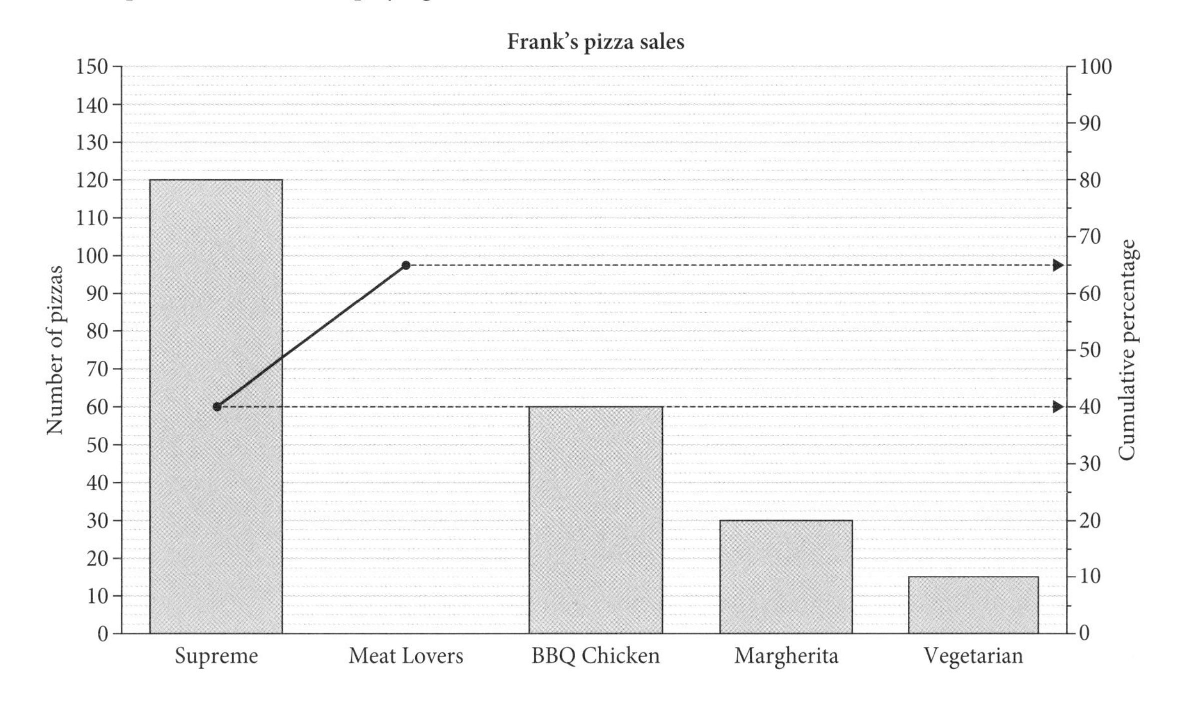

a Complete the bar chart by drawing in the column for 'Meat Lovers'. 1 mark

b Complete the cumulative frequency line graph. 1 mark

c Frank wants to find the median of the data. Explain briefly why this is not possible. 1 mark

Question 5 (2 marks)

A set of examination marks is normally distributed with a mean of 66.

Anlong's mark of 60 in the examination converts to a z-score of $-\frac{2}{3}$.

Find the standard deviation of the data set. 2 marks

Question 6 (8 marks)

Ben records the heights of all of the Year 12 students at his school, correct to the nearest centimetre.

He has represented his results in the cumulative frequency histogram below.

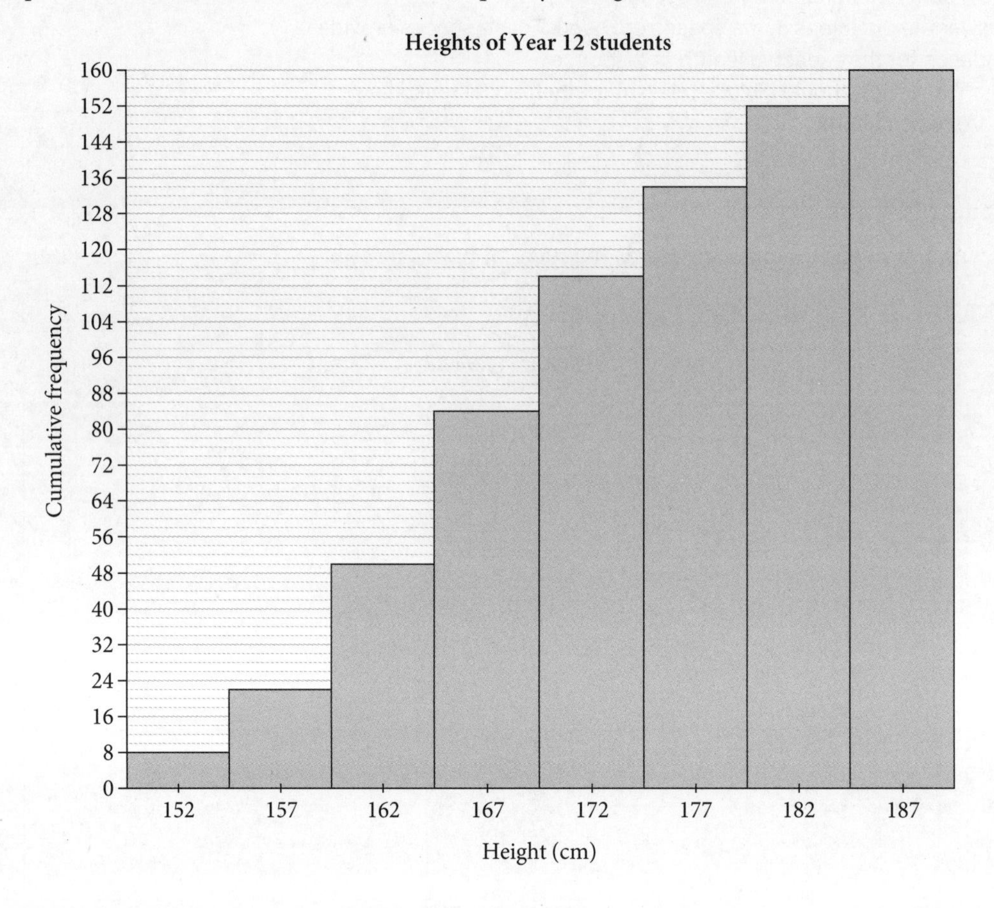

a On the above graph, draw a cumulative frequency polygon. 1 mark

b How can you tell that Ben has grouped his data into classes? 1 mark

c Find an estimate for:

i the median. 1 mark

ii the interquartile range. 2 marks

d The lowest and highest heights in Ben's data are 152 cm and 188 cm respectively.

In the space below, draw a box plot that summarises Ben's data. 3 marks

Questions 4–6 are worth 13 marks in total (Section II halfway point)

Question 7 (6 marks)

Packets of brown sugar are produced by a facility in Queensland. The masses of the packets are known to be normally distributed with a mean of 1.06 kg and a standard deviation of 0.03 kg.

a Find the z-score that corresponds to a mass of 1.05 kg. 1 mark

b A certain packet is known to have a z-score of 2.

Find its mass. 1 mark

c Approximately what percentage of sugar packets have a mass of between 1.09 kg and 1.15 kg? 2 marks

d The labelling on the packets states that the mass is 1 kg.

If a sample of 2600 packets are examined, how many would be expected to contain less than the stated mass of 1 kg? 2 marks

Question 8 (2 marks)

A set of data has a lower quartile of 36 and an upper quartile of 48.5.

Determine whether a score of 20 would be considered an outlier. 2 marks

Question 9 (9 marks)

Madeleine is a geneticist. She is undertaking a study to predict the height (y cm) of a child at age 18 years based on the average height (x cm) of the parents.

The table shows the average heights of 15 sets of parents and the corresponding heights of their 18-year-old children. One child's height has been omitted and is represented by A.

x	168	159	165	170	172	169	171	175	168	176	173	156	154	162	166
y	165	161	160	169	175	173	164	170	172	180	A	164	158	169	168

a Identify in words the independent variable in Madeleine's data set. 1 mark

b The scatterplot represents the data in the table but the point (169, 173) has been omitted.

Complete the scatterplot by representing this entry with a cross. 1 mark

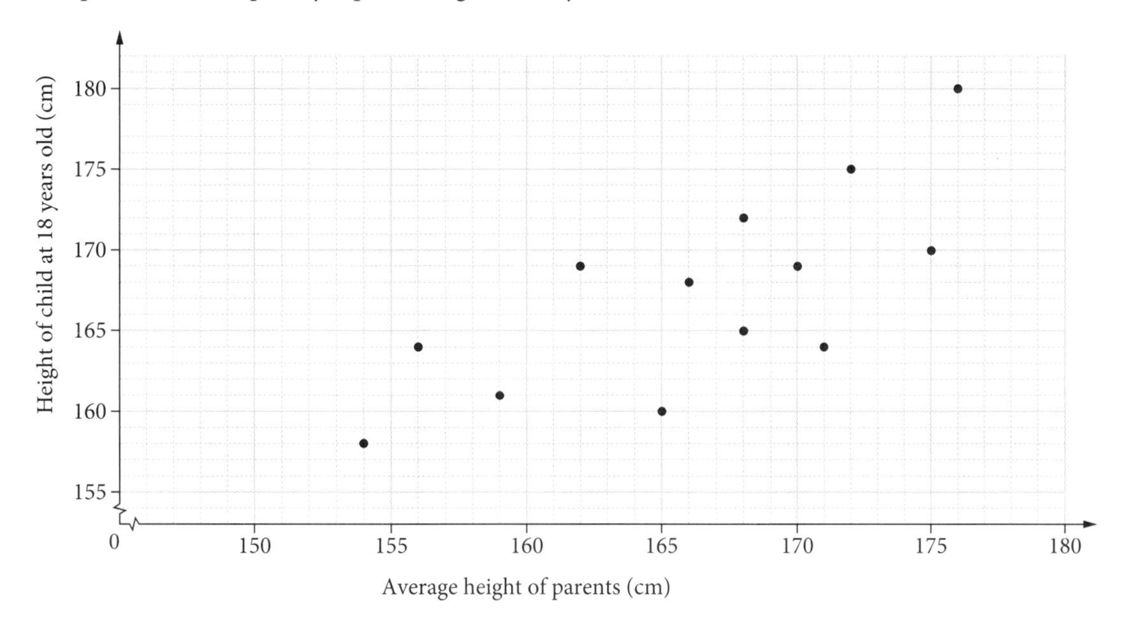

c Using your calculator, determine correct to four decimal places Pearson's correlation coefficient for the 14 pairs of measurements available. 1 mark

d Using your calculator, find the equation of the least-squares regression line, writing the gradient and y-intercept to three decimal places. 2 marks

e Use your equation in part **d** to find an estimated value for A in the table from part **a**. Round your answer to one decimal place. 1 mark

f Comment briefly on the reliability of your answer in part **e**. 1 mark

g LeBron and Savannah have an average height of 190 cm. Use the least-squares regression line to predict the height of their daughter at age 18.

Would such a prediction be valid? Give reasons for your answer. 2 marks

END OF PAPER

WORKED SOLUTIONS

Section I (1 mark each)

Question 1

B 0 0 1 1 | 1 2 3 3 | 4 5 5 5 | 6 6 7 8

Q_1 = 1, Median = 3.5, Q_3 = 5.5

So median = 3.5 and IQR = 5.5 − 1 = 4.5.

Straightforward question. Common content with Maths Standard 2.

Question 2

C As 40 < 50, the mean has been reduced.
40 is more than 1 standard deviation below the mean (10 > 7), so the spread has increased causing the standard deviation to increase.

Interpretation question. Common content with Maths Standard 2.

Question 3

C The correlation coefficient of −0.92 implies that the 'dots' will closely approximate a straight line with a gradient of −1. The scatterplot that does this best is plot C.

Straightforward question. Common content with Maths Standard 2.

Section II (✓ = 1 mark)

Question 4 (3 marks)

a, b The column for Supreme pizza shows 120 people ordered Supreme.

This corresponds to 40% of sales.

So 10% of sales = 120 ÷ 4 = 30 pizzas.
So 100% of sales = 30 × 10 = 300 pizzas.

65% of sales come from Supreme and Meat Lovers combined, so 65% − 40% = 25% of sales can be attributed to Meat Lovers.

25% of 300 = 75

So, height of Meat Lovers column is 75.

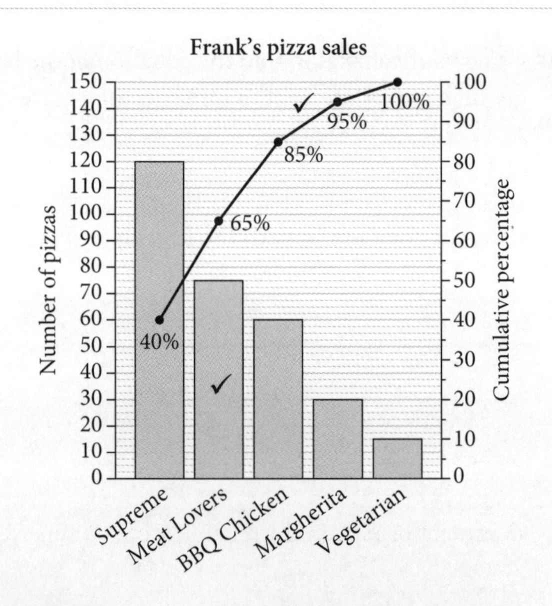

Category	Frequency	Cumulative frequency	Cumulative percentage
Supreme	120	120	40%
Meat Lovers	75	195	65%
BBQ Chicken	60	255	$\frac{255}{300} \times 100\% = 85\%$
Margherita	30	285	$\frac{285}{300} \times 100\% = 95\%$
Vegetarian	15	300	100%

Pareto charts are new to both the Maths Advanced and Standard 2 courses.
Creating a table makes constructing the graph much easier.
Remember that the scales are on opposite sides of the Pareto chart.

c Because the data is categorical and cannot be ordered, it is not possible to find the median. The only relevant measure of central tendency for categorical data is the mode. ✓

Understand more than just how to find the mean, median and mode. This question tests deeper conceptual understanding of the measures of central tendency.

Question 5 (2 marks)

$$z = \frac{x - \mu}{\sigma}$$

$$-\frac{2}{3} = \frac{60 - 66}{\sigma} \checkmark$$

$$-\frac{2}{3} = -\frac{6}{\sigma}$$

$$\frac{2}{3} = \frac{6}{\sigma}$$

$$2\sigma = 18$$

$$\sigma = 9 \checkmark$$

Straightforward question. Common content with Standard 2. The z-score formula is on the HSC exam reference sheet.

Question 6 (8 marks)

a

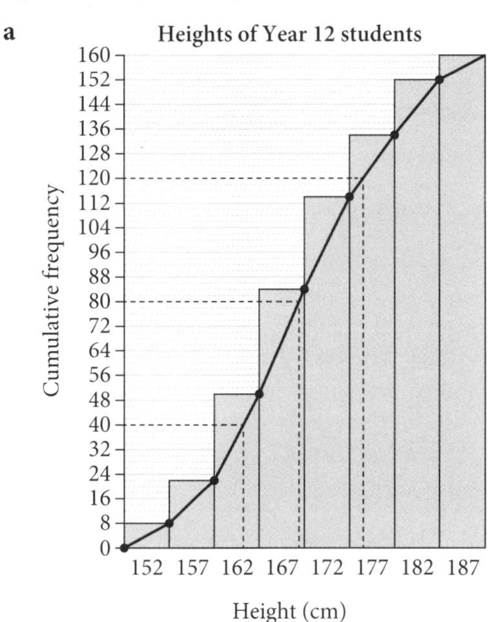

✓ for correctly placed polygon.

Cumulative frequency graphs are common content with Maths Standard 2.

b The numbers labelling the columns on the horizontal axis are not consecutive, so they must be class centres representing classes; for example, 162 represents 160 to 164 cm. ✓

This is another question that tests deeper understanding and details that students often fail to notice.

c i 160 students, so for the median draw dotted line from $\frac{1}{2} \times 160 = 80$ on the cumulative frequency axis and read the matching value on the heights axis.

Median ≈ 169 cm ✓

ii 160 students, so for Q_1 draw dotted line from $\frac{1}{4} \times 160 = 40$ on the cumulative frequency axis and read the matching value on the heights axis.

$Q_1 = 163$ cm ✓

For Q_3 draw a dotted line from $\frac{3}{4} \times 160 = 120$ on the cumulative frequency axis and read the matching value on the heights axis.

$Q_3 = 175$ cm

IQR ≈ 175 − 163 = 12 cm ✓

Common content with Maths Standard 2.

d Five-number summary: 152, 163, 169, 175, 188

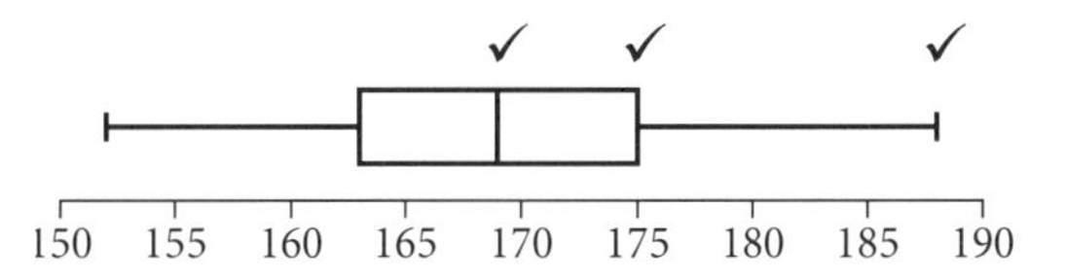

Straightforward question. Don't forget to include a scale.

Question 7 (6 marks)

a
$$z = \frac{x - \mu}{\sigma}$$

$$z = \frac{1.05 - 1.06}{0.03}$$

$$= \frac{-0.01}{0.03} \checkmark$$

$$= -\frac{1}{3}$$

b Rearranging the formula for z-scores to make x the subject,

$$x = \mu + z\sigma$$
$$= 1.06 + 2(0.03)$$
$$= 1.12\text{ kg} \checkmark$$

Parts **a** and **b** involve basic formulas and equations. Common content with Maths Standard 2.

c

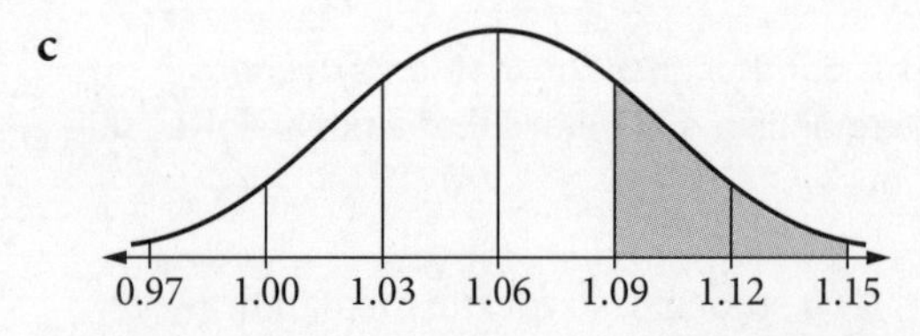

According to the empirical rule, approximate percentage of scores in the shaded region

$= \dfrac{99.7 - 68}{2}$ ✓

$= 15.85\%$ ✓

Hint

Instead of using the empirical rule, some students like to memorise the areas of the individual regions to make calculations easier in problems. The area from 0 to 1 is 34% ($\frac{1}{2}$ of 68%), then moving right it's 13.5%, 2.35% and 0.15%.

It is the same going from 0 to −1 and so on. So the answer to part **c** is 13.5 + 2.35 = 15.85%.

Very common question style. Students must be able to manipulate the 3 basic percentages given in the empirical rule. Common content with Maths Standard 2.

d

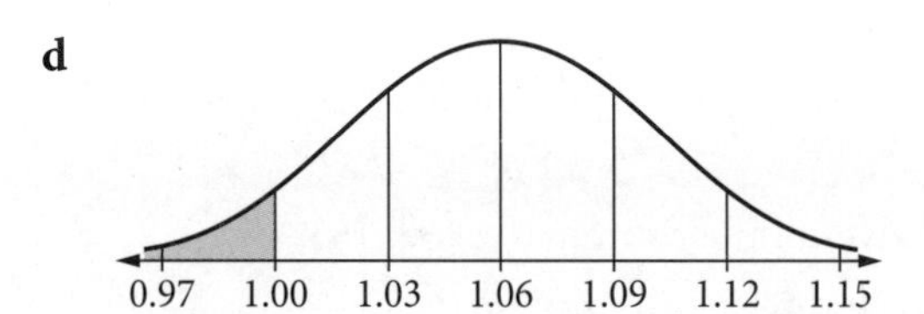

Percentage of scores in the shaded region

$\dfrac{100 - 95}{2} = 2.5\%$. ✓

So number of packets = 2.5% × 2600 = 65. ✓

Common content with Maths Standard 2.

Question 8 (2 marks)

To be considered a low outlier,

data value $< Q_1 - 1.5 \times \text{IQR}$

$\text{IQR} = 48.5 - 36 = 12.5$ ✓

$Q_1 - 1.5 \times \text{IQR} = 36 - 1.5 \times 12.5$

$= 17.25$

So 20 would not be considered an outlier because it is not less than 17.25. ✓

This formula appears on the HSC exam reference sheet.

Question 9 (9 marks)

a The independent variable is the average height of the parents. ✓

Know your definitions. Dependent and independent variables are specifically mentioned in the syllabus.

b

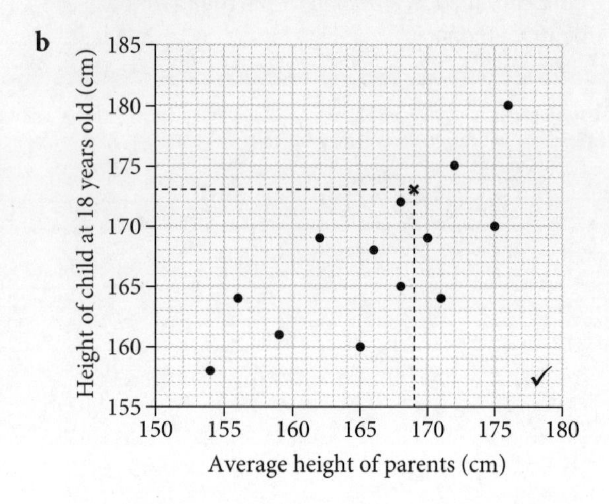

Straightforward question. Common content with Maths Standard 2.

c $r \approx 0.7397$ ✓

d $y = 0.681x$ ✓ $+ 54.328$ ✓

e $y = 0.681x + 54.328$

When $x = 173$,

$y = 0.681(173) + 54.328$ ✓

≈ 172.1 cm

Question **9** parts **c**, **d** and **e** are all straightforward questions.

f Since this is an interpolation (that is, 173 lies within the data set), there is a high degree of reliability. ✓

g $y = 0.681x + 54.328$

When $x = 190$,

$y = 0.681(190) + 54.328$

≈ 183.7 cm ✓

However, since in this case we are extrapolating (that is, using an x-value outside of the existing data), we cannot be as confident in our prediction.

The model may not hold for such values. ✓

Distinguishing between interpolation and extrapolation and the limitations of each is specifically mentioned in the syllabus.

HSC exam topic grid (2011–2020)

This table shows the coverage of this topic in past HSC exams by question number. The past exams can be downloaded from the NESA website (www.educationstandards.nsw.edu.au) by selecting 'Year 11 – Year 12', 'HSC exam papers'. NESA marking feedback and guidelines can also be found there.

Statistics and bivariate data were introduced to the Mathematics Advanced course in 2020, but both are common content with Mathematics Standard 2. In the table below, * refers to past HSC questions (2011–2019) in Mathematics Standard 2 and Mathematics General 2. Before 2019, 'Mathematics Standard 2' was called 'Mathematics General 2', and before 2014, 'General Mathematics'. For these exams, select 'Year 11 – Year 12', 'Resources archive', 'HSC exam papers archive'.

	Displaying and analysing data	**Comparing data**	**Scatterplots and correlation**	**Linear regression**
2011*	7, 14, 17, 25(a)(i), 25(d)	11, 25(b)	8	
2012*	1, 26(e)	28(d)	11, 29(a)	19 (replace 'median regression' with 'line of best fit')
2013*	6, 14, 15, 26(b), 27(c), 29(b)(i)	26(f)	2	28(b)
2014*	14, 26(e), 30(b)(ii)–(iv)	29(c)	30(b)(i)	30(b)(vi)–(viii)
2015*	4, 6, 27(d), 29(d)		28(e)	28(e)
2016*	7, 19, 21, 27(c)	22, 29(c)	3, 29(d)(i)	29(d)(ii) (use calculator and table of values)
2017*	1, 27(a), 29(d), 30(a)		12	
2018*	1, 3, 6, 11, 26(e)	26(d)		29(d) (use calculator and table of values for (i))
2019*	10, 19, 39	39	23(a)–(b)	23(c)
2020 new course	27			27

See the topic grid in Chapter 7, p. 88 for the normal distribution and z-scores.

CHAPTER 7 TOPIC EXAM

7

Probability distributions

MA-S3 Random variables

S3.1 Continuous random variables

Note: This exam also tests Year 11 probability. However, z-scores and the normal distribution are covered in the Statistics and bivariate data topic exam in Chapter 6.

- A reference sheet is provided on page 195 at the back of this book
- For questions in Section II, show relevant mathematical reasoning and/or calculations

Reading time: 4 minutes
Working time: 1 hour
Total marks: 33

Section I – 3 questions, 3 marks
- Attempt Questions 1–3
- Allow about 5 minutes for this section

Section II – 7 questions, 30 marks
- Attempt Questions 4–10
- Allow about 55 minutes for this section

Section I

- Attempt Questions 1–3 **3 marks**
- Allow about 5 minutes for this section

Question 1

The probability that a soccer player will score when taking a penalty kick is $\frac{5}{7}$.

If he takes two penalty kicks, what is the probability that he doesn't score on both?

A $\frac{4}{49}$

B $\frac{10}{49}$

C $\frac{25}{49}$

D $\frac{39}{49}$

Question 2

Consider the Venn diagram on the right. An element from the sample space is selected at random.

What is the value of $P(A)$?

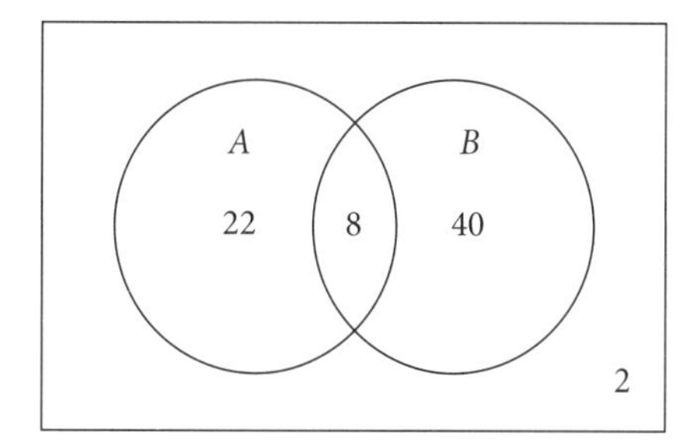

A $\frac{11}{36}$

B $\frac{11}{35}$

C $\frac{5}{12}$

D $\frac{3}{7}$

Question 3

A bag contains 10 red, 20 yellow and x green lollies. Demi chooses a green lolly and eats it.

If she selects a second lolly at random, what is the probability that it is also green?

A $\frac{x}{29 + x}$

B $\frac{x - 1}{30 + x}$

C $\frac{x - 1}{29 + x}$

D $\frac{x}{30 + x}$

Section II

- Attempt Questions 4–10 **30 marks**
- Allow about 55 minutes for this section
- Answer the questions in the spaces provided. These spaces provide guidance for the expected length of response.
- Your responses should include relevant mathematical reasoning and/or calculations.

Question 4 (5 marks)

The spinner below is divided into three sectors of equal area. They are coloured red, yellow and green, as shown.

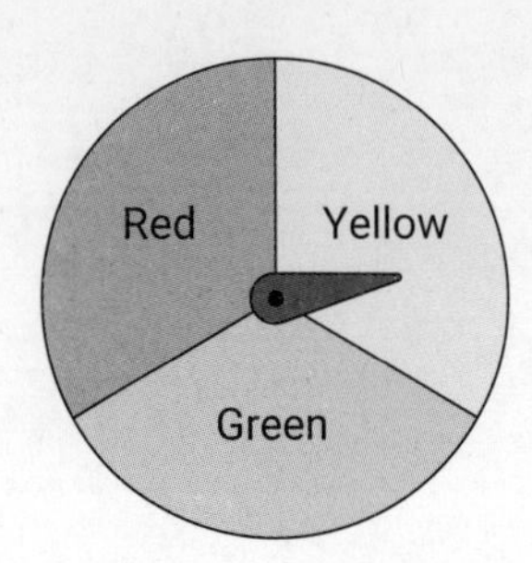

William spins the spinner 5 times. Let the random variable, X, be the number of times the spinner lands on red.

This discrete random variable has the following probability distribution:

x	0	1	2	3	4	5
$P(X = x)$	$\frac{32}{243}$	k	k	$\frac{40}{243}$	$\frac{10}{243}$	$\frac{1}{243}$

a Find the value of k. 2 marks

b Find $E(X)$. 1 mark

c Find the standard deviation of the distribution as a surd. 2 marks

Question 5 (4 marks)

A box of chocolates contains 20 chocolates. There are 12 with a caramel centre (C) and 8 with a mint centre (M). Kailesh chooses 2 chocolates and eats them.

a Complete the probability tree diagram below by placing probabilities on the remaining branches. 2 marks

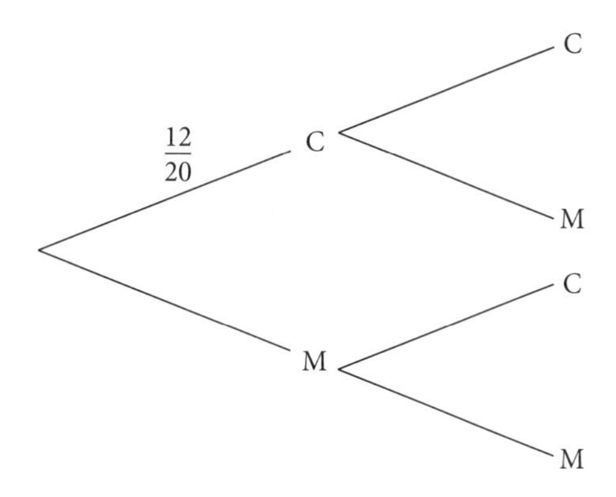

b What is the probability that Kailesh chooses 2 chocolates with the same centre? 2 marks

Question 6 (2 marks)

A blue die and a red die are rolled. Both dice show the numbers 1 to 6.

If the blue die shows an even number, find the probability that the sum of the numbers on both dice is greater than 8. 2 marks

TOPIC EXAM

Question 7 (5 marks)

This incomplete two-way table shows the proportion by age and sex of the patrons in a cinema.

	Male	Female	Total
Child	0.35		0.62
Adult			
Total	0.42		1

a Explain why the value in the shaded cell is 1. 1 mark

b Complete the table. 1 mark

c Find the probability that a randomly-selected patron from this cinema:

i is female. 1 mark

ii is an adult male. 1 mark

d Find the probability that a randomly selected male is a child. 1 mark

Questions 4–7 are worth 16 marks in total (Section II halfway point)

TOPIC EXAM

Question 8 (4 marks) ©NESA 2019 HSC EXAM, QUESTION 15(d)

The probability that a person chosen at random has red hair is 0.02.

a Two people are chosen at random.

What is the probability of at least ONE having red hair? 2 marks

b What is the least number of people that can be chosen at random so that the probability of at least ONE person with red hair is greater than 0.4? 2 marks

Question 9 (4 marks) ©NESA 2018 HSC EXAM, QUESTION 16(b)

A game involves rolling two six-sided dice, followed by rolling a third six-sided die. To win the game, the number rolled on the third die must lie between the two numbers rolled previously. For example, if the first two dice show 1 and 4, the game can only be won by rolling a 2 or 3 with the third die.

a What is the probability of a player having no chance of winning before rolling the third die? 2 marks

b What is the probability of a player winning the game? 2 marks

Question 10 (6 marks)

X is a continuous random variable that is defined by the probability density function

$$f(x) = \begin{cases} k(x-1)^2 & 2 \leq x \leq 4 \\ 0 & \text{otherwise} \end{cases}$$

a Show that $k = \dfrac{3}{26}$. 2 marks

b Find the cumulative distribution function. 2 marks

c Show that the upper quartile is approximately 3.748. 2 marks

END OF PAPER

WORKED SOLUTIONS

Section I (1 mark each)

Question 1

A $P(\text{player scores}) = \frac{5}{7}$

$P(\text{player doesn't score}) = \frac{2}{7}$

So P(player doesn't score in 2 shots)

$= \frac{2}{7} \times \frac{2}{7} = \frac{4}{49}$.

Straightforward probability question.

Question

C Number of elements in set $A = 22 + 8 = 30$

Number of elements in set $S = 22 + 8 + 40 + 2$
$= 72$

So $P(A) = \frac{30}{72} = \frac{5}{12}$.

Straightforward question. See also the 2020 HSC exam, Question 14. Venn diagrams are specifically mentioned in the syllabus.

Question 3

C Originally, the bag contains x green lollies and $(30 + x)$ lollies in total.

Upon selection of the first lolly, which we know is green, there are now $(x - 1)$ green lollies and $(29 + x)$ lollies in total.

So $P(\text{second lolly is green}) = \frac{x - 1}{29 + x}$.

Harder question. Generalised algebraic example of a familiar scenario.

Section II (✓ = 1 mark)

Question 4 (5 marks)

a $\frac{32}{243} + k + k + \frac{40}{243} + \frac{10}{243} + \frac{1}{243} = 1$ ✓

$$2k + \frac{83}{243} = 1$$

$$2k = \frac{160}{243}$$

$$k = \frac{80}{243}$$ ✓

Discrete probability distributions is a new topic. The sum of all probabilities must be 1.

b $E(X) = 0\left(\frac{32}{243}\right) + 1\left(\frac{80}{243}\right) + 2\left(\frac{80}{243}\right) + 3\left(\frac{40}{243}\right) + 4\left(\frac{10}{243}\right) + 5\left(\frac{1}{243}\right)$

$= \frac{405}{243}$

$= \frac{5}{3}$ ✓

Know how to find $E(X)$ or μ. The formula $E(X) = \mu = \Sigma x \times p(x)$ is not on the HSC exam reference sheet.

c $\text{Var}(X) = E(X^2) - \mu^2$

$$= 0\left(\frac{32}{243}\right) + 1\left(\frac{80}{243}\right) + 4\left(\frac{80}{243}\right) + 9\left(\frac{40}{243}\right) + 16\left(\frac{10}{243}\right) + 25\left(\frac{1}{243}\right) - \left(\frac{5}{3}\right)^2 \checkmark$$

$$= \frac{10}{9}$$

So $\sigma = \sqrt{\frac{10}{9}} = \frac{\sqrt{10}}{3}$. ✓

This formula is on the HSC exam reference sheet in *both* its forms: $\text{Var}(X) = E\left[(X - \mu)^2\right] = E(X^2) - \mu^2$
The second form as used in the solution is easier to use. Remember that standard deviation $\sigma = \sqrt{\text{Var}(X)}$.

Question 5 (4 marks)

a

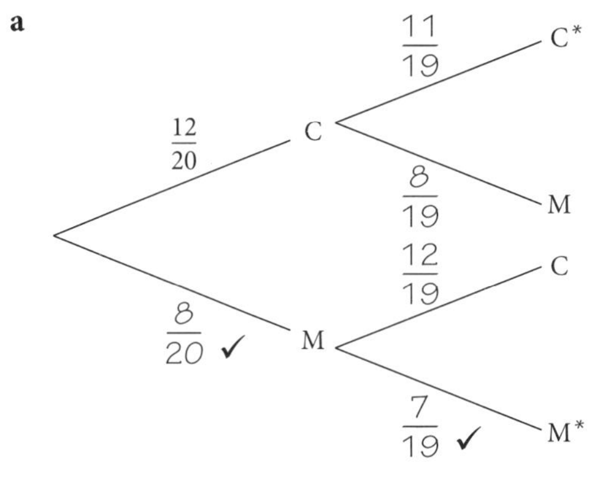

Typical tree diagram question. See also 2008 HSC exam, Question 7(c).

b The branches in the probability tree diagram that satisfy the condition 'have the same centre' are marked with *.

$$P(\text{same centre}) = \left(\frac{12}{20} \times \frac{11}{19}\right) \checkmark + \left(\frac{12}{20} \times \frac{11}{19}\right) \checkmark$$

$$= \frac{47}{95}$$

Common HSC exam question. Be aware of events that can happen in more than one way.

Question 6 (2 marks)

The sample space for rolling 2 dice is shown below. However, the results where the blue die is odd have been omitted, as the question tells us 'the blue die is even'.

		Blue die					
		1	**2**	**3**	**4**	**5**	**6**
Red die	**1**	–	3	–	5	–	7
	2	–	4	–	6	–	8
	3	–	5	–	7	–	**9**
	4	–	6	–	8	–	**10**
	5	–	7	–	**9**	–	**11**
	6	–	8	–	**10**	–	**12** ✓

The results with a sum greater than 8 are in bold.

So $P(> 8 \mid \text{blue die is even}) = \frac{6}{18} = \frac{1}{3}$. ✓

Conditional probability is new to the course, introduced in 2020. Conditional probability reduces the number of possible (and favourable) outcomes.

Question 7 (5 marks)

a The shaded cell shows 1 because the sum of the proportions of all possible patrons must be 1. ✓

The events 'patron is male' and 'patron is female' are complementary, so the sum of their probabilities is 1. Similarly, the events 'patron is an adult' and 'patron is a child' are also complementary.

Understand the underlying concepts of probability.

b

	Male	Female	Total
Child	0.35	0.27	0.62
Adult	0.07	0.31	0.38
Total	0.42	0.58	1

✓

c **i** $P(\text{female}) = 0.58$ ✓

ii $P(\text{adult male}) = 0.07$ ✓

Parts **b** and **c** are both straightforward questions.

d $\left(\text{child}\middle|\text{male}\right) = \dfrac{0.35}{0.42} = \dfrac{5}{6}$ ✓

Conditional probability using proportion.

Note that we do not know how many patrons are in the cinema, but this is not necessary.

Question 8 (4 marks)

a $P(\text{not red hair}) = 1 - 0.02$
$= 0.98$ ✓

If the selection is done twice,

$P(\text{at least one has red hair}) = 1 - P(\text{no red hair})$
$= 1 - 0.98^2$ ✓
$= 0.0396$

2019 HSC exam question. Questions using the phrase 'at least one' are very common in HSC exam papers. Remember, 'at least one' = 1 – 'none'.

b If $P(\text{at least one with red hair}) = 1 - 0.98^2$ when the selection is done twice, $P(\text{at least one with red hair})$ when the selection is done n times will be $1 - 0.98^n$.

We require $1 - 0.98^n > 0.4$. ✓

$$0.6 > 0.98^n$$
$$\ln(0.6) > \ln(0.98^n)$$
$$\ln(0.6) > n\ln(0.98)$$
$$\frac{\ln(0.6)}{\ln(0.98)} < n \quad \text{reversing inequality because } \ln(0.98) \text{ is negative}$$
$$n > 25.285\ldots$$

So, a minimum of 26 people must be chosen for the probability of at least one of them having red hair is greater than 0.4. ✓

A Band 6 question. Generalisation of the idea in part **a**, and also the use of logarithms to solve an exponential inequality. Common student errors were not noting that ln (0.98) was negative and not rounding *up* the solution to the inequality.

Question 9 (4 marks)

a Recall all of the 36 possible outcomes when two dice are rolled.

	1	2	3	4	5	6
1	**1,1**	**1,2**	1,3	1,4	1,5	1,6
2	**2,1**	**2,2**	**2,3**	1,5	2,5	2,6
3	3,1	**3,2**	**3,3**	**3,5**	3,5	3,6
4	4,1	4,2	**4,3**	**4,5**	**4,5**	4,6
5	5,1	5,2	5,3	**5,4**	**5,5**	**5,6**
6	6,1	6,2	6,3	6,4	**6,5**	**6,6**

✓

To have no chance of winning before rolling the third die, when the first two are rolled, the numbers must be either the same or consecutive (differ by 1). These are shown in bold above.

So $P(\text{no chance of winning}) = \dfrac{16}{36} = \dfrac{4}{9}$. ✓

Aimed at Band 6 students, this 2018 HSC exam question has been included here for its uniqueness.

This unusual question would have tested students for deep understanding and adaptability to think outside the square.

b We must consider all of the cases separately.

Case 1:

$P(\text{the numbers on the two dice differ by 2}) = \dfrac{8}{36}$

$P(\text{the third number is between the first two}) = \dfrac{1}{6}$

So $P(\text{Case 1 occurs}) = \dfrac{8}{36} \times \dfrac{1}{6} = \dfrac{8}{216}$.

Case 2:

$P(\text{the numbers on the two dice differ by 3}) = \dfrac{6}{36}$

$P(\text{the third number is between the first two}) = \dfrac{2}{6}$

So $P(\text{Case 2 occurs}) = \dfrac{6}{36} \times \dfrac{2}{6} = \dfrac{12}{216}$.

Case 3:

$P(\text{the numbers on the two dice differ by 4}) = \dfrac{4}{36}$

$P(\text{the third number is between the first two}) = \dfrac{3}{6}$

So $P(\text{Case 3 occurs}) = \dfrac{4}{36} \times \dfrac{3}{6} = \dfrac{12}{216}$.

Case 4:

$P(\text{the numbers on the two dice differ by 5}) = \frac{2}{36}$

$P(\text{the third number is between the first two}) = \frac{4}{6}$

So $P(\text{Case 4 occurs}) = \frac{2}{36} \times \frac{4}{6} = \frac{8}{216} + 1.$

✓ for identifying one possible case.

$$P(\text{winning the game}) = \frac{8}{216} + \frac{12}{216} + \frac{12}{216} + \frac{8}{216}$$
$$= \frac{40}{216}$$
$$= \frac{5}{27} \checkmark$$

Another atypical high-level question with different possible methods of solution. You will need to have a deep understanding to successfully respond to questions like this.

Question 10 (6 marks)

a
$$\int_2^4 k(x-1)^2\,dx = 1 \checkmark$$
$$\left[\frac{k(x-1)^3}{3}\right]_2^4 = 1$$
$$\frac{k(4-1)^3}{3} - \frac{k(2-1)^3}{3} = 1$$
$$\frac{k}{3}(3^3 - 1^3) = 1$$
$$\frac{26k}{3} = 1$$

So $k = \frac{3}{26}$, as required. ✓

Continuous probability distributions are new to the course. This question examines the property of PDFs that state $\int_{-\infty}^{\infty} f(x)dx = 1$.

See also the 2020 HSC exam, Question 23(a) and sample 2020 HSC exam, Question 31 (in the A+ Mathematics Advanced Study Notes book also).

b
$$\text{CDF} = \int_2^x \frac{3}{26}(x-1)^2\,dx \checkmark$$
$$= \left[\frac{(x-1)^3}{26}\right]_2^x$$
$$= \frac{(x-1)^3}{26} - \frac{1}{26}$$
$$= \frac{(x-1)^3 - 1}{26} \checkmark$$

Be able to determine a cumulative density function by integrating its associated probability density function.

c Upper quartile is found by solving CDF = 0.75.

$$\frac{(x-1)^3 - 1}{26} = 0.75 \checkmark$$
$$(x-1)^3 - 1 = 19.5$$
$$(x-1)^3 = 20.5$$
$$x - 1 = \sqrt[3]{20.75}$$
$$\text{So } x = \sqrt[3]{20.75} + 1$$
$$= 3.7479\ldots \checkmark$$
$$\approx 3.748, \text{ as required.}$$

Know how to find the median or any percentile (in this case, the 75th) using the CDF.

HSC exam topic grid (2011–2020)

This table shows the coverage of this topic in past HSC exams by question number. The past exams can be downloaded from the NESA website (www.educationstandards.nsw.edu.au) by selecting 'Year 11 – Year 12', 'HSC exam papers'. NESA marking feedback and guidelines can also be found there.

Probability distributions were introduced to the Mathematics Advanced course in 2020. The normal distribution and z-scores are common content with Mathematics Standard 2. In the table below, * refers to past HSC questions (2011–2019) in Mathematics Standard 2 (2019) and Mathematics General 2. Before 2019, 'Mathematics Standard 2' was called 'Mathematics General 2', and before 2014, 'General Mathematics'. For these exams, select 'Year 11 – Year 12', 'Resources archive', 'HSC exam papers archive'.

	Discrete random variables (Year 11)	**Continuous random variables**	**The normal distribution**	**z-scores**
2011*				27(c)
2012*			29(b)	
2013*			20	29(b)
2014*			24	
2015*			20	28(b)
2016*			13	
2017*			29(d)	13
2018*			23	27(e)
2019*			15	38
2020 new course		23	9, 28	3

Note: The normal distribution and z-scores are covered in the topic exam of Chapter 6.

Mathematics Advanced

PRACTICE MINI-HSC EXAM 1

General instructions

- Reading time: 4 minutes
- Working time: 1 hour
- A reference sheet is provided on page 195 at the back of this book
- For questions in Section II, show relevant mathematical reasoning and/or calculations

Total marks: 33

Section I – 3 questions, 3 marks

- Attempt Questions 1–3
- Allow about 5 minutes for this section

Section II – 10 questions, 30 marks

- Attempt Questions 4–13
- Allow about 55 minutes for this section

Section I

3 marks
Attempt Questions 1–3
Allow about 5 minutes for this section.

Circle the correct answer.

Question 1

What are the solutions to the quadratic equation $3x^2 - x - 4 = 0$?

A $x = -1$ and $x = \frac{4}{3}$

B $x = -1$ and $x = -\frac{4}{3}$

C $x = 1$ and $x = \frac{4}{3}$

D $x = 1$ and $x = -\frac{4}{3}$

Question 2

Consider the triangle ABC below.

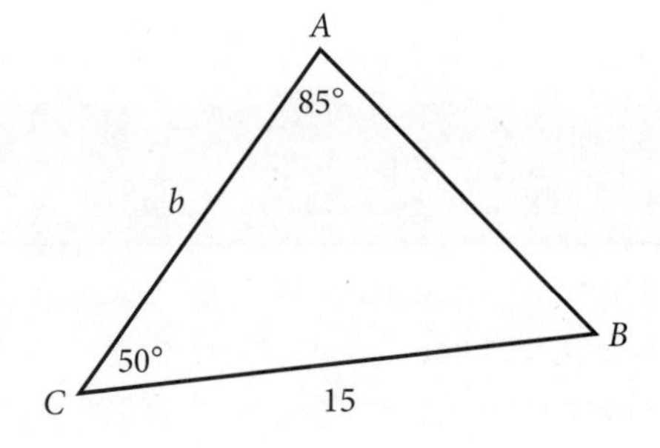

Which expression gives the value of b?

A $\dfrac{15\sin 50°}{\sin 85°}$

B $\dfrac{15\sin 85°}{\sin 50°}$

C $\dfrac{15\sin 45°}{\sin 85°}$

D $\dfrac{15\sin 85°}{\sin 45°}$

Question 3

The displacement of a particle moving in a straight line is given by the equation $x = 2t - \frac{1}{2}t^2$, where x is in metres, t is in seconds and $t \geq 0$.

Which of the following statements is FALSE?

A The particle is at the origin after 4 seconds.

B The particle is initially moving to the left.

C The acceleration of the particle is constant.

D The speed of the particle increases indefinitely.

Section II

30 marks
Attempt Questions 4–13
Allow about 55 minutes for this section.

- Answer the questions in the spaces provided. These spaces provide guidance for the expected length of response.
- Your responses should include relevant mathematical reasoning and/or calculations.

Question 4 (2 marks)

Solve $|3x - 2| = 5$. 2 marks

Question 5 (2 marks)

Evaluate $\int_0^{\frac{\pi}{4}} 2\cos 2x \, dx$. 2 marks

Question 6 (2 marks)

AJ invests \$12 000 into an account earning 6% p.a. with interest compounded monthly for a period of 5 years.

Calculate the amount of interest AJ earns, correct to the nearest dollar. 2 marks

Question 7 (2 marks)

Calculate the exact area of the sector below, centred at O. 2 marks

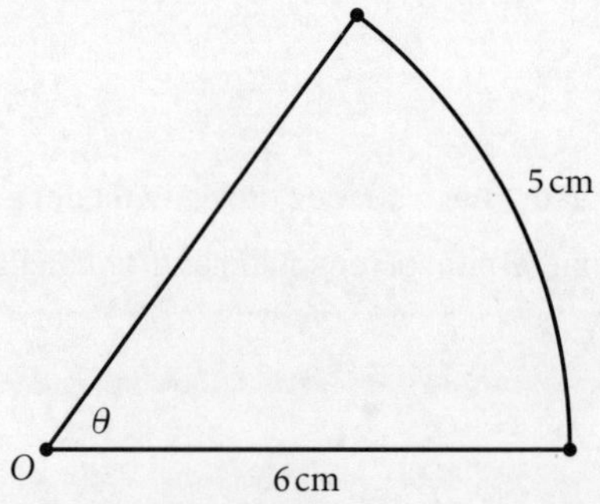

Question 8 (3 marks)

Solve $2\sin\theta - \operatorname{cosec}\theta = 1$ for θ in the domain $[0, 2\pi]$. 3 marks

Question 9 (2 marks)

Find $\int \frac{3x}{2x^2+1} dx$. 2 marks

Questions 4–9 are worth 13 marks in total (Section II halfway point)

Question 10 (7 marks)

The function $f(x) = \frac{x^3}{6} - \frac{x^2}{4} - 3x + 1$ is defined for x in the domain $[-4, 6]$.

a Find the stationary points and determine their nature. 3 marks

Question 10 continues on page 94

Question 10 (continued)

b Find the global maximum value of $f(x)$. 2 marks

c Sketch $y = f(x)$, showing the stationary points and both endpoints. 2 marks

End of Question 10

Question 11 (3 marks)

A bag contains some black marbles and some white marbles.

Ryan draws three marbles from the bag. The discrete random variable, X, represents the number of black marbles drawn.

The experiment is repeated 100 times and the relative frequencies appear in the table below.

X	0	1	2	3
Relative frequency	0.12	0.22	0.46	k

a Find k. 1 mark

b Find μ. 1 mark

c Find $\mathrm{Var}(X)$. 1 mark

Question 12 (5 marks)

A machine produces metal rods used to reinforce concrete slabs. The diameters of the rods are normally distributed with a mean of 8 mm and a standard deviation of 0.3 mm.

a Find the diameter of a rod with a z-score of 1.2. 1 mark

b If a rod is selected at random, find the probability that the diameter of the rod is less than 7.7 mm. 2 marks

c If 2 rods are selected at random, find the probability that exactly 1 of the rods will have a diameter that is less than 7.7 mm. 2 marks

9780170459235

Question 13 (2 marks)

In ΔXYZ, $XY = 14$ cm, $XZ = 12$ cm and $YZ = 16$ cm.

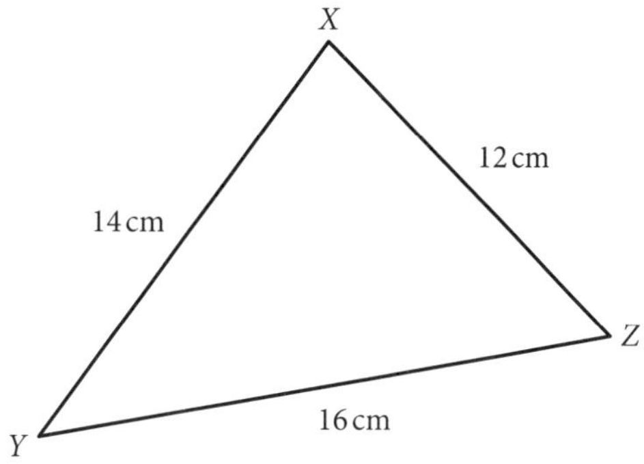

X

14 cm

12 cm

Z

Y

16 cm

a Show that $\angle Y \approx 47°$. 1 mark

b Find the area of ΔXYZ, to the nearest cm^2. 1 mark

END OF PAPER

WORKED SOLUTIONS

Section I (1 mark each)

Question 1

A $3x^2 - x - 4 = 0$

$(3x - 4)(x + 1) = 0$

So $x = -1$ and $\frac{4}{3}$.

Straightforward question.

Question 2

C $\angle B = 180° - 85° - 50° = 45°$

By the sine rule, $\frac{b}{\sin B} = \frac{a}{\sin A}$

$\frac{b}{\sin 45°} = \frac{15}{\sin 85°}$.

$b = \frac{15 \sin 45°}{\sin 85°}$

Common HSC exam question. The sine rule is common content with Maths Standard 2.

Question 3

B Check A: $x = 2t - \frac{1}{2}t^2$

When $t = 4$,

$x = 2(4) - \frac{1}{2}(4)^2 = 0$ True

Check B: $x = 2t - \frac{1}{2}t^2$ so $v = 2 - t$

When $t = 0$, $v = 2 - 0 = 2 > 0$.

This means the particle is initially moving right, making statement B false.

Check C: As $v = 2 - t$, $a = 2$. True

Check D: $v = 2 - t$, means it keeps decreasing, so its speed keeps increasing. True

Be able to interpret your answers when applying calculus to motion in a straight line.

Section II (✓ = 1 mark)

Question 4 (2 marks)

$|3x - 2| = 5$

So $3x - 2 = 5$ OR $3x - 2 = -5$

$3x = 7$ $\quad$ $3x = -3$

$x = \frac{7}{3}$ ✓ $\quad$ $x = -1$ ✓

Straightforward question on absolute value equations which commonly appear in exams.

Question 5 (2 marks)

$$\int_0^{\frac{\pi}{4}} 2\cos 2x\, dx = [\sin 2x]_0^{\frac{\pi}{4}} \checkmark$$

$$= \sin\left(\frac{\pi}{2}\right) - \sin 0$$

$$= 1 \checkmark$$

Trigonometric integrals are very common in the HSC exam.

Question 6 (2 marks)

$r = 6\%$ p.a. $= 0.5\%$ per month $= 0.005$

$n = 60$ ✓

$A = P(1 + r)^n$

$= 12\,000(1 + 0.005)^{60}$

$\approx \$16\,186$ ✓

So interest = \$16 186 − \$12 000 = \$4186.

Straightforward compound interest question. Common content with Maths Standard 2. This formula is on the HSC exam reference sheet.

Question 7 (2 marks)

$l = r\theta$

$5 = 6\theta$

So $\theta = \frac{5}{6}$. ✓

$A = \frac{1}{2}r^2\theta$

$= \frac{1}{2} \times 6^2 \times \frac{5}{6}$

$= 15\text{ cm}^2$ ✓

Unusual question using formulas that appear on the reference sheet. Need to work backwards to find θ first.

Question 8 (3 marks)

$$2\sin\theta - \operatorname{cosec}\theta = 1$$

$$2\sin\theta - \frac{1}{\sin\theta} = 1$$

$$2\sin^2\theta - 1 = \sin\theta$$

$$2\sin^2\theta - \sin\theta - 1 = 0 \checkmark$$

$$(2\sin\theta + 1)(\sin\theta - 1) = 0$$

So $\sin\theta = -\frac{1}{2}$ OR $\sin\theta = 1$

$\theta = \frac{7\pi}{6}, \frac{11\pi}{6}$ ✓ $\qquad \theta = \frac{\pi}{2}$ ✓

Solving this type of trigonometric equation is a higher-order skill but one that students should be familiar with. Practise these question types.

Question 9 (2 marks)

$$\int \frac{3x}{2x^2+1}dx = 3\int \frac{x}{2x^2+1}dx$$

$$= \frac{3}{4}\int \frac{4x}{2x^2+1}dx \checkmark$$

$$= \frac{3}{4}\ln(2x^2+1) + c \checkmark$$

Straightforward question. Always look out for integrals of the form $\int \frac{f'(x)}{f(x)}dx$ that become logarithmic functions and don't forget '+ c'.

Question 10 (7 marks)

a $f(x) = \frac{x^3}{6} - \frac{x^2}{4} - 3x + 1$

So $f'(x) = \frac{x^2}{2} - \frac{x}{2} - 3.$ ✓

Stationary points occur when $f'(x) = 0$.

$$0 = \frac{x^2}{2} - \frac{x}{2} - 3$$

$$0 = x^2 - x - 6$$

$$0 = (x-3)(x+2)$$

So $x = -2$ and 3.

$$f(-2) = \frac{(-2)^3}{6} - \frac{(-2)^2}{4} - 3(-2) + 1$$

$$= -\frac{4}{3} - 1 + 6 + 1$$

$$= 4\frac{2}{3}$$

$$f(3) = \frac{(3)^3}{6} - \frac{(3)^2}{4} - 3(3) + 1$$

$$= \frac{27}{6} - \frac{9}{4} - 9 + 1$$

$$= -5\frac{3}{4} \checkmark$$

Use $f''(x)$ to determine nature.

$$f''(x) = x - \frac{1}{2}$$

$$f''(-2) = -2 - \frac{1}{2}$$

$$= -2\frac{1}{2}$$

$$< 0$$

So a local maximum occurs at $\left(-2, 4\frac{2}{3}\right)$.

$$f''(x) = x - \frac{1}{2}$$

$$f''(3) = 3 - \frac{1}{2}$$

$$= 2\frac{1}{2}$$

$$> 0$$

So a local minimum occurs at $\left(4, -5\frac{3}{4}\right)$. ✓

Curve sketching is in almost all HSC exams and is usually worth 3–4 marks due to the many steps involved. Take care when substituting. This is a common area where students will make careless mistakes.

b Check the endpoints of the domain and compare with the maximum point above for global maximum.

$$f(-4) = \frac{(-4)^3}{6} - \frac{(-4)^2}{4} - 3(-4) + 1$$

$$= -10\frac{2}{3} - 4 + 12 + 1$$

$$= -1\frac{2}{3} \checkmark$$

$$f(6) = \frac{(6)^3}{6} - \frac{(6)^2}{4} - 3(6) + 1$$

$$= 36 - 9 - 18 + 1$$

$$= 10$$

$$> 4\frac{2}{3}$$

So the global maximum of $f(x)$ is 10. ✓

Be familiar with domain restrictions.

c

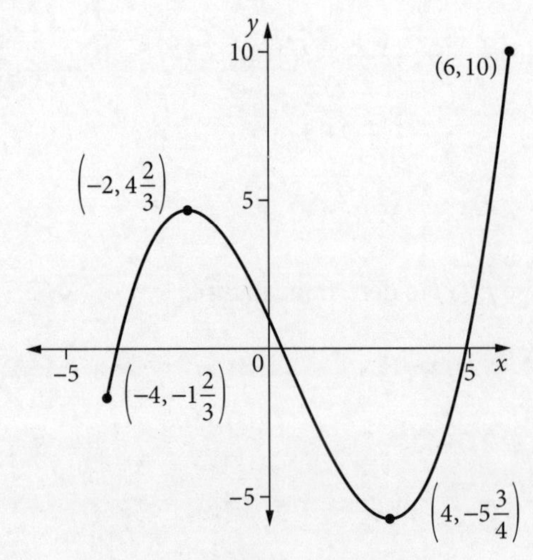

✓ stationary points and shape

✓ endpoints

Draw smooth curves and rule your axes. Accuracy is an important consideration in Mathematics.

Question 11 (3 marks)

a $k + 0.12 + 0.22 + 0.46 = 1$

So $k = 0.2$. ✓

b $\mu = \Sigma x \times \text{relative frequency}$

$= (0 \times 0.12) + (1 \times 0.22) + (2 \times 0.46) + (3 \times 0.2)$

$= 1.74$ ✓

Discrete probability is new content. Try to do as many questions on this topic as possible.

c $E(X^2) = (0 \times 0.12) + (1 \times 0.22) + (4 \times 0.46) + (9 \times 0.2)$

$= 3.86$

$\text{Var}(X) = E(X^2) - \mu^2$

$= 3.86 - 1.74^2$

$= 0.8324$ ✓

Know how to find the variance. It is a relatively simple process that with practise should become straightforward.

Question 12 (5 marks)

a Rewrite the z-score formula with x as the subject:

$$z = \frac{x - \mu}{\sigma}$$

$x = z\sigma + \mu$

$= 1.2(0.3) + 8$

$= 8.36\text{ mm}$ ✓

Straightforward question. Common content with Maths Standard 2.

b $z = \frac{x - \mu}{\sigma}$

$= \frac{7.7 - 8}{0.3}$

$= -1$ ✓

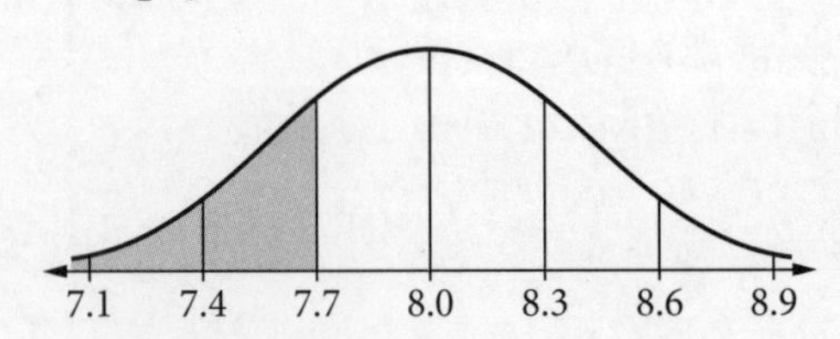

According to the empirical rule, approximately 68% of rods will have a diameter between 7.7 mm and 8.3 mm.

Therefore, 100% − 68% = 32% of diameters will be below 7.7 mm or above 8.3 mm.

Due to the symmetrical nature of the normal distribution,

P(diameter of rod is less than 7.7 mm)

$= \frac{0.32}{2} = 0.16$ ✓

The empirical rule is on the reference sheet and is common content with Maths Standard 2. Practise calculating areas under the normal curve. An alternative solution is 0.15% + 2.35% + 13.5% = 16%.

c For exactly 1 of the 2 rods to have a diameter that is less than 7.7 mm, this rod may either be selected first or second (and the other diameter is greater than 7.7 mm).

Consider both cases and note that $P(x > 7.7) = 1 - 0.16 = 0.84$. ✓

P(exactly 1 rod's diameter is less than 7.7 mm)

$= (0.16 \times 0.84) + (0.84 \times 0.16)$

$= 0.2688$ ✓

Draw a tree diagram if it helps. HSC exam questions such as this can be achieved in more than one way.

Question 13 (2 marks)

a $\cos Y = \frac{14^2 + 16^2 - 12^2}{2 \times 14 \times 16}$

$\angle Y = 46.5674\ldots$

$\approx 47°$, as required. ✓

b $\text{Area} = \frac{1}{2}xz\sin Y$

$= \frac{1}{2} \times 14 \times 16 \times \sin 47°$

$\approx 81\text{ cm}^2$ ✓

Note: the answer to part **a** is required for part **b**. Typical HSC exam question. Common content with Maths Standard 2.

Mathematics Advanced

PRACTICE MINI-HSC EXAM 2

General instructions	• Reading time: 4 minutes • Working time: 1 hour • A reference sheet is provided on page 195 at the back of this book • For questions in Section II, show relevant mathematical reasoning and/or calculations
Total marks: 33	**Section I – 3 questions, 3 marks** • Attempt Questions 1–3 • Allow about 5 minutes for this section **Section II – 9 questions, 30 marks** • Attempt Questions 4–12 • Allow about 55 minutes for this section

Section I

3 marks
Attempt Questions 1–3
Allow about 5 minutes for this section

Circle the correct answer.

Question 1

What is the x-intercept of the line with equation $2x + 3y + 5 = 0$?

A $-\frac{5}{2}$

B $-\frac{5}{3}$

C $\frac{5}{3}$

D $\frac{5}{2}$

Question 2

What is the period of the function $y = \tan 4x$?

A $\frac{\pi}{4}$

B $\frac{\pi}{2}$

C 2π

D 4π

Question 3

Find $\int e^{2x+1}\, dx$.

A $2e^{2x+1} + c$

B $\frac{e^{2x+2}}{2} + c$

C $(2x + 1)e^{2x} + c$

D $\frac{e^{2x+1}}{2} + c$

Section II

30 marks
Attempt Questions 4–12
Allow about 55 minutes for this section.

- Answer the questions in the spaces provided. These spaces provide guidance for the expected length of response.
- Your responses should include relevant mathematical reasoning and/or calculations.

Question 4 (2 marks)

Find $\int \frac{1}{(x+4)^2}\,dx$. 2 marks

Question 5 (2 marks)

Consider the function $f(x) = 2x^2 - 3x + 2$. Show that $f(x) > 0$ for all x. 2 marks

Question 6 (4 marks)

Peta has a bank account with a balance of $300. She begins a plan to save $30 per week from her part-time job.

Peta's balance, P, is modelled by the equation $P = 300 + 30t$, where t is the number of weeks since her savings plan started.

a On the grid below, draw the graph of this model. 1 mark

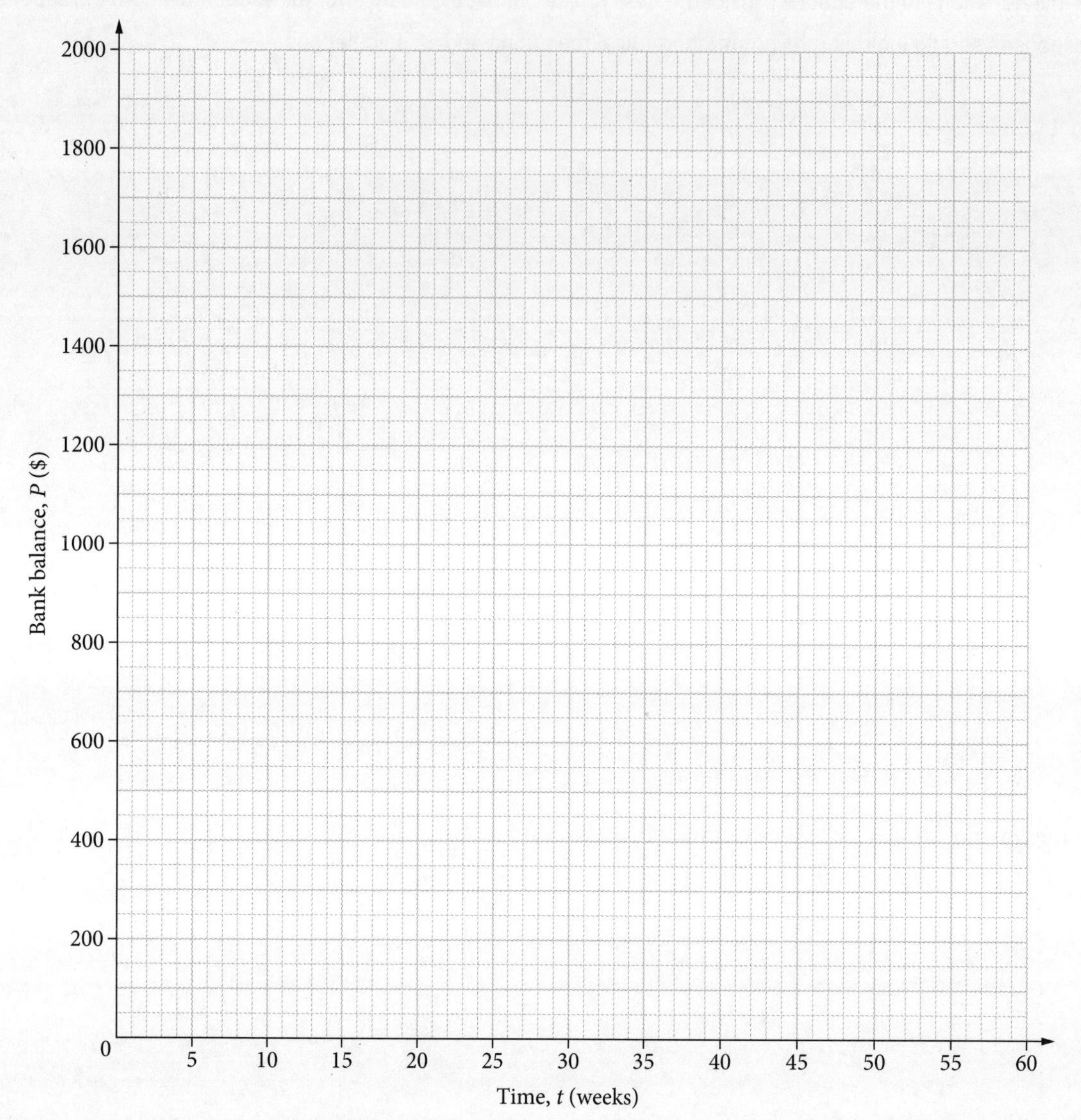

b Peta's sister, Rita, begins saving 10 weeks after Peta. Rita saves $50 per week.

By drawing a line on the grid in part **a**, or otherwise, find when the balances in Peta's and Rita's accounts are equal. 2 marks

c Using the graphs drawn, or otherwise, find when their combined balance is \$1400. 1 mark

Question 7 (2 marks)

Differentiate $f(x) = 2x^2 - x$ from first principles using the formula 2 marks

$$f'(x) = \lim_{h \to 0} \frac{f(x+h) - f(x)}{h}.$$

Question 8 (6 marks)

A particle, initially at the origin, is moving in a straight line. Its velocity is given by $v = 2 - 2t$ m/s.

a Find the initial velocity of the particle. 1 mark

b Find when the particle is next at the origin. 2 marks

Question 8 continues on page 106

Question 8 (continued)

c Find when the particle is at rest. 1 mark

d Find the distance travelled by the particle in the first 4 seconds. 2 marks

Questions 4–8 are worth 16 marks in total (Section II halfway point)

Question 9 (6 marks)

A closed cylinder is constructed using exactly $1000\,\text{cm}^2$ of aluminium.

a Show that the volume, $V\,\text{cm}^3$, of the cylinder is given by $V = 500r - \pi r^3$. 3 marks

b Find the value of r that gives the maximum volume of the cylinder. 3 marks

Question 10 (3 marks)

A set of exam results is normally distributed with a mean of 64 and a standard deviation of 10.

a Edmond scores 84 in the exam. Express this as a z-score. 1 mark

b Two students are chosen at random. What is the probability that both students score higher than 84? 2 marks

Question 11 (2 marks)

Use the table of values and the trapezoidal rule with 5 function values to approximate $\int_0^2 f(x)\,dx$. 2 marks

x	0	$\frac{1}{2}$	1	$1\frac{1}{2}$	2
$f(x)$	3	$4\frac{1}{2}$	$6\frac{1}{2}$	9	7

Question 12 (3 marks)

Consider the infinite geometric series $3 + \frac{3\sqrt{2}}{4} + \frac{3}{8} + \ldots$

a Explain why this series has a limiting sum. 2 marks

b Find the exact sum of the infinite series. 1 mark

END OF PAPER

WORKED SOLUTIONS

Section I (1 mark each)

Question 1

A $2x + 3y + 5 = 0$

When $y = 0$,

$$2x + 5 = 0$$
$$2x = -5$$
$$x = -\frac{5}{2}.$$

Calculating intercepts on a linear graph were studied in Years 9–10.

Question 2

A Period of $y = \tan bx$ is $\frac{\pi}{b}$

So the period of $y = \tan 4x$ is $\frac{\pi}{4}$.

Straightforward question. Know the formula for the period of a tangent function (it is not on the HSC exam reference sheet).

Question 3

D $\int e^{2x+1}\,dx = \frac{e^{2x+1}}{2} + c$

Application of the result

$$\int f'(x)e^{f(x)}\,dx = e^{f(x)} + c$$

This formula is on the HSC exam reference sheet.

Section II (✓ = 1 mark)

Question 4 (2 marks)

$$\int \frac{1}{(x+4)^2}\,dx = \int (x+4)^{-2}\,dx \quad ✓$$
$$= \frac{(x+4)^{-1}}{-1} + C \quad ✓$$
$$= \frac{-1}{x+4} + C$$

Typical integration question involving negative powers. Often asked to see which students are able to distinguish between integrals that produce logarithmic functions and those that do not.

Question 5 (2 marks)

$f(x) = 2x^2 - 3x + 2$

This function represents a parabola which is concave up since $a = 2 > 0$. ✓

Check the discriminant, Δ.

$$\Delta = b^2 - 4ac$$
$$= (-3)^2 - 4 \times 2 \times 2$$
$$= -7 \quad ✓$$

As $\Delta < 0, f(x) > 0$ for all x, as required.

This question requires consideration of the graph of $y = f(x)$ although it is not mentioned. There is also no mention of the discriminant. An alternative method is to show that the vertex is above the x-axis. It is up to the student to know it is required.

Question 6 (4 marks)

a, b Rita's graph should begin at $(10, 0)$ and have a gradient of 50, so another point on the line is $(20, 50 \times 10 = 500)$.

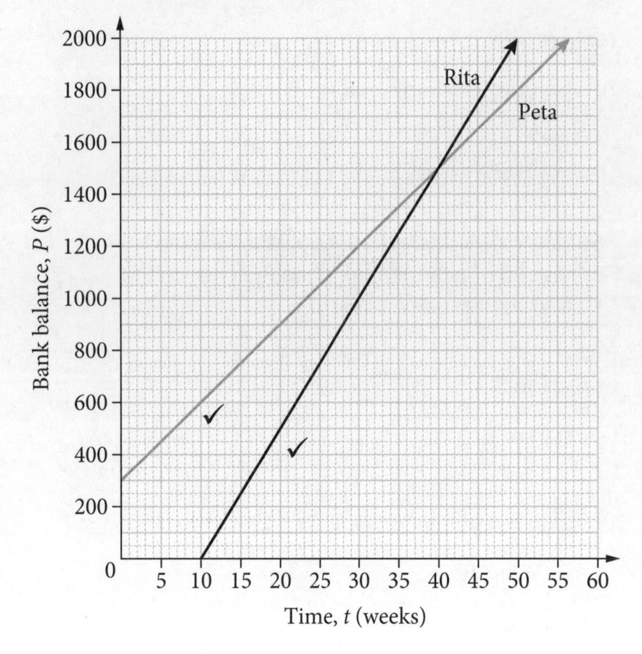

Graphically, the point of intersection occurs when $t = 40$, so the balances are equal after 40 weeks. ✓

Applying and interpreting intersection of lines.
Common content with Maths Standard 2.
A similar question appeared in the 2020 HSC exam, Question 11.

c Looking at the 2 graphs, we need a sum of \$1400. Look for when the average of the 2 balances is \$700. When $t = 20$, Rita's balance is \$500 and Peta's balance is \$900, so their sum equals \$1400.

This problem can also be solved algebraically. Peta's savings is $P = 300 + 30t$.

Rita is saving \$50 per week, so her savings will be $R = 50t + c$. Substituting the point $(10, 0)$ which is on the line, $c = -500$. Therefore, $R = 50t - 500$.

So we require,

$$\begin{aligned} 300 + 30t + 50t - 500 &= 1400 \\ 80t - 200 &= 1400 \\ 80t &= 1600 \\ t &= 20 \text{ weeks} \checkmark \end{aligned}$$

Although the question does not specifically ask for the equation that models Rita's savings, it is required if you want to solve the simultaneous equations algebraically.

Question 7 (2 marks)

$$f'(x) = \lim_{h \to 0} \frac{f(x+h) - f(x)}{h}$$

Given that $f(x) = 2x^2 - x$,

$$\begin{aligned} f(x+h) &= 2(x+h)^2 - (x+h) \\ &= 2(x^2 + 2xh + h^2) - (x+h) \\ &= 2x^2 + 4xh + 2h^2 - x - h \checkmark \end{aligned}$$

$$\begin{aligned} f'(x) &= \lim_{h \to 0} \frac{2x^2 + 4xh + 2h^2 - x - h - (2x^2 - x)}{h} \\ &= \lim_{h \to 0} \frac{4xh + 2h^2 - h}{h} \\ &= \lim_{h \to 0} 4x + 2h - 1 \\ &= 4x - 1 \checkmark \end{aligned}$$

Differentiation from first principles is a higher-order skill but one that students should be familiar with.
Practise these question types.

Question 8 (6 marks)

a $v = 2 - 2(0)$
$= 2\text{ m/s}$ ✓

b $v = 2 - 2t$

So $x = 2t - t^2 + c$. ✓

When $t = 0$, $x = 0$, so $c = 0$.

$$\begin{aligned} x &= 2t - t^2 \\ 0 &= 2t - t^2 \\ 0 &= t(2 - t) \end{aligned}$$

So $t = 2\text{ s}$ $(t \neq 0)$. ✓

c $0 = 2 - 2t$
So $t = 1\text{ s}$. ✓

Parts **a** to **c** are straightforward questions. Applying calculus to motion questions is very common in HSC examinations.

d When $t = 0, x = 0$.

This is given in the initial conditions.

When $t = 1, x = 2(1) - 1^2$
$= 1\text{ m}$ ✓

Examine this as this is when the particle comes to rest.

When $t = 4, x = 2(4) - 1^2$
$= -8\text{ m}$

So the particle starts at the origin, moves right to $x = 1$ and then travels left to $x = -8$.

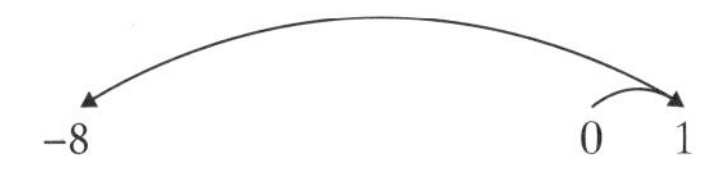

So the particle travels a total of 10 m in the first 4 seconds. ✓

Note that part **c** is leading into part **d**. The fact the particle stops between $t = 0$ and $t = 4$ should guide you to approach the question as has been done in the solution.

Question 9 (6 marks)

a
$$A = 2\pi r^2 + 2\pi rh$$
$$1000 = 2\pi r^2 + 2\pi rh \ ✓$$
$$500 = \pi r^2 + \pi rh$$
$$\pi rh = 500 - \pi r^2$$
$$\text{So } h = \frac{500}{\pi r} - \frac{\pi r^2}{\pi r}.$$
$$h = \frac{500}{\pi r} - r \ [*] \ ✓$$

Now $V = \pi r^2 h$
$$= \pi r^2\left(\frac{500}{\pi r} - r\right), \text{ from } [*]$$
$$= \left(\frac{500\pi r^2}{\pi r} - \pi r^3\right)$$

So $V = 500r - \pi r^3$, as required. ✓

Maximisation and minimisation questions can be challenging.

They often appear towards the end of HSC examinations and target students aiming for a Band 6 standard of achievement.

Note that while the surface area formula for a cylinder is on the HSC exam reference sheet, the required volume formula is not.

b $V = 500r - \pi r^3$

So $\dfrac{dV}{dr} = 500 - 3\pi r^2$. ✓

V is maximised when $\dfrac{dV}{dr} = 0$.

$$0 = 500 - 3\pi r^2$$
$$3\pi r^2 = 500$$
$$r^2 = \frac{500}{3\pi}$$

So $r = \sqrt{\dfrac{500}{3\pi}}$ cm, $r > 0$. ✓

You cannot stop here even though the question tells you that you are seeking a maximum and there is only one value for r. You *must* prove that this value maximises V and does not minimise it.

$$\frac{d^2V}{dr^2} = -6\pi r^2$$
$$r^2 = \frac{500}{3\pi}$$
$$= -6\pi\left(\frac{500}{3\pi}\right)$$
$$= -1000$$
$$< 0$$

So V is maximised when $r = \sqrt{\dfrac{500}{3\pi}}$ cm. ✓

Even if you were unable to establish the result in part **a**, you should be able to attempt part **b** given the question was phrased as a 'show that' question.

Question 10 (3 marks)

a
$$z = \frac{x - \mu}{\sigma}$$
$$= \frac{84 - 64}{10}$$
$$= 2 \ ✓$$

Straightforward question. Common content with Maths Standard 2.

b $P(z > 2) = \dfrac{1 - 0.95}{2}$
$= 0.025$ ✓

So $P(\text{both score over } 84) = 0.025^2$
$= 0.000\,625$ ✓

Remember that 95% of scores lie within 2 standard deviations of the mean.

Question 11 (2 marks)

$\int f(x)dx$

$= \frac{h}{2}\left[f(0) + 2\left(f\left(\frac{1}{2}\right) + f(1) + f\left(1\frac{1}{2}\right)\right) + f(2)\right]$

$= \frac{\frac{1}{2}}{2}\left[3 + 2\left(4\frac{1}{2} + 6\frac{1}{2} + 9\right) + 7\right]$ ✓

$= \frac{1}{4}[3 + 40 + 7]$

$= 12.5$ ✓

Straightforward application of the trapezoidal rule.

Question 12 (3 marks)

a $r = T_2 \div T_1$

$= 3 \div \frac{3\sqrt{2}}{4}$

$= \frac{\sqrt{2}}{4}$ ✓

$= 0.3535\ldots$

As $-1 < r < 1$, the series has a limiting sum. ✓

For full marks, you must know the condition for a geometric series to have a limiting sum.

b $S = \frac{a}{1 - r}$

$= \frac{3}{1 - \frac{\sqrt{2}}{4}}$

$= \frac{3}{\frac{4 - \sqrt{2}}{4}}$

$= \frac{12}{4 - \sqrt{2}}$ [*] ✓

$= \frac{12}{4 - \sqrt{2}} \times \frac{4 + \sqrt{2}}{4 + \sqrt{2}}$

$= \frac{48 + 12\sqrt{2}}{14}$

$= \frac{24 + 6\sqrt{2}}{7}$

The extensive simplification is not required. Stopping at the line marked [*] is acceptable. Do not, however, approximate the answer with a calculator. Especially when the question specifically asks for the *exact* answer.

Mathematics Advanced

PRACTICE HSC EXAM 1

General instructions

- Reading time: 10 minutes
- Working time: 3 hours
- A reference sheet is provided on page 195 at the back of this book
- For questions in Section II, show relevant mathematical reasoning and/or calculations

Total marks: 100

Section I – 10 questions, 10 marks

- Attempt Questions 1–10
- Allow about 15 minutes for this section

Section II – 21 questions, 90 marks

- Attempt Questions 11–31
- Allow about 2 hours and 45 minutes for this section

Section I

10 marks
Attempt Questions 1–10
Allow about 15 minutes for this section

Circle the correct answer.

Question 1

What is $\sqrt{112}$ in its simplest form?

A $2\sqrt{28}$

B $4\sqrt{7}$

C $2\sqrt{12}$

D $11\sqrt{2}$

Question 2

Between which 2 consecutive numbers is the value of $\log_3 100$?

A 0 and 1

B 1 and 2

C 4 and 5

D 33 and 34

Question 3

The scatterplot below shows the relationship between two quantities x and y.

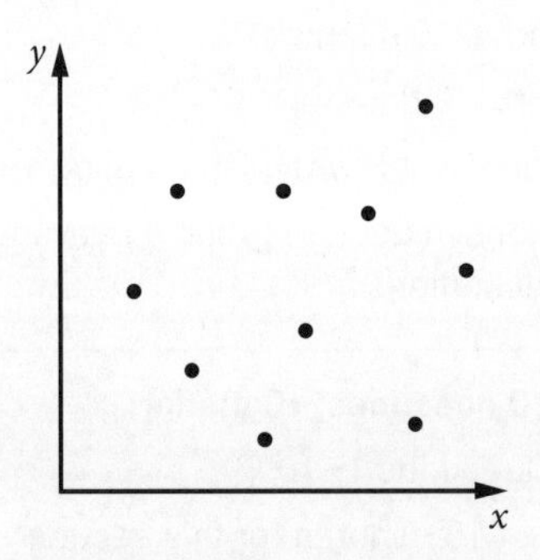

Which of the following values is closest to Pearson's correlation coefficient for the data in the scatterplot?

A −0.95

B −0.1

C 0.9

D 1

Question 4

For a particular function $y = f(x)$:

$$f(3) = -6$$

$$f'(3) = 0$$

$$f''(3) = 4$$

Which of the following statements is true regarding the graph of $y = f(x)$?

A It passes through the point $(3, 0)$.

B There is a local minimum at $(3, -6)$.

C It is concave down at $(3, -6)$.

D There is a point of inflection at $(3, 4)$.

Question 5

Simplify $\dfrac{\operatorname{cosec}\theta}{\cot\theta}$.

A $\sec\theta$

B $\sin\theta$

C $\tan\theta\cos\theta$

D $\tan\theta\sin\theta$

Question 6

A number of people were tested for a virus; however, the test is not always correct. The results appear in the table below. A positive test result is recorded when the test indicates the person has the virus.

	Test is positive	Test is negative	Total
People with the virus	394	86	480
People without the virus	30	420	450
Total	424	506	930

One test result is selected at random.

What is the probability that the test result is correct?

A 46%

B 82%

C 88%

D 93%

Question 7

The heights of a population of people are normally distributed. It is known that $P(z < a) = 0.7$, where $a > 0$.

What is the value of $P(-1 < z < a)$?

A 0.4

B 0.54

C 0.68

D 0.95

9780170459235

Question 8

Find $\frac{d}{dx}\int x^3\,dx$.

A $3x^2$

B $\frac{x^4}{4}$

C $\frac{x^4}{4}-\frac{1}{4}$

D x^3

Question 9

How many solutions does the equation $e^x(\cos x+1)=0$ have for $0 \le x \le 2\pi$?

A 0

B 1

C 2

D 3

Question 10

The graph of the function $y=f(x)$ has a minimum point at $P(-4,-6)$.

What is the corresponding maximum point on the graph of $y=-f(2x)$?

A $(-4,6)$

B $(-8,6)$

C $(-2,6)$

D $(2,-6)$

Section II

90 marks
Attempt Questions 11–31
Allow about 2 hours and 45 minutes for this section

- Answer the questions in the spaces provided. These spaces provide guidance for the expected length of response.
- Your responses should include relevant mathematical reasoning and/or calculations.

Question 11 (4 marks)

The equation $y = \frac{2}{3}x + 4$ represents a straight line, as shown.

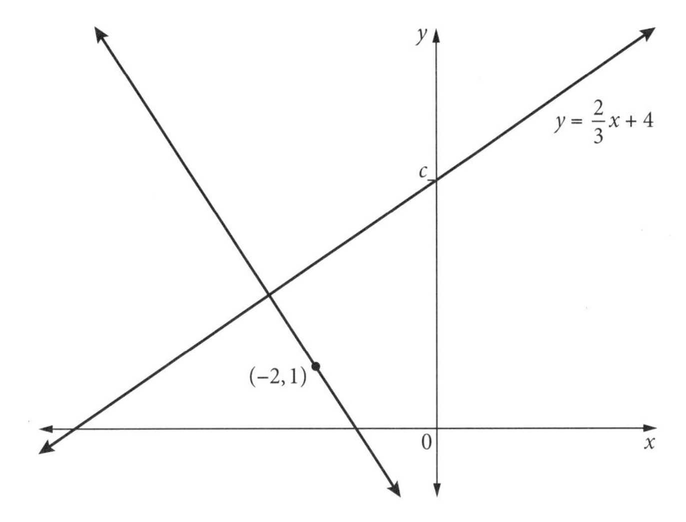

a What is the geometrical significance of the coefficient of x in the equation? 1 mark

b The line of $y = \frac{2}{3}x + 4$ crosses the y-axis at c.

What is the value of c? 1 mark

c A line passing through the point $(-2, 1)$ is perpendicular to the line $y = \frac{2}{3}x + 4$, as shown above.

Find the equation of this line. 2 marks

Question 12 (2 marks)

Find $\frac{d}{dx}\left(\frac{x^2}{x+1}\right)$. 2 marks

Question 13 (2 marks)

Solve this pair of simultaneous equations. 2 marks

$$2x - 2y = 3$$
$$x + 4y = 9$$

Question 14 (3 marks)

Evaluate $\int_0^{\frac{\pi}{3}} \cos 2x \, dx$. 3 marks

Question 15 (5 marks)

Sketch the graph of $y = x^3 - 4x^2$. Label the stationary points, the point of inflection and the x- and y-intercepts of the graph. 5 marks

Question 16 (5 marks)

A survey was taken of 100 people who were asked if they had a subscription to two different online streaming services, Netflex (N) and Span (S).

The results showed that:

- 15 people did not subscribe to either service
- 65 people subscribed to Netflex
- 50 people subscribed to Span

a Use the results to complete the Venn diagram below. 1 mark

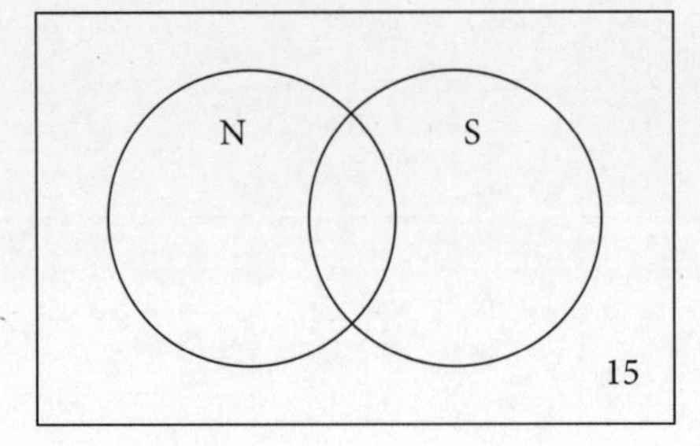

b One of the people surveyed is selected at random.

Find the probability that they subscribe to both streaming services. 1 mark

c A randomly-selected person is known to subscribe to a streaming service.

Find the probability that this person subscribes to Span (S). 1 mark

d Two people are chosen at random.

Find the probability that exactly one of them is a subscriber to a streaming service. 2 marks

Question 17 (3 marks)

The first term of an arithmetic sequence is 8 and the 15th term is 57.

a Find the common difference, d. 1 mark

b Find S_{20}, the sum of the first 20 terms. 2 marks

Question 18 (3 marks)

a Find the centre and radius of the circle $x^2 - 8x + y^2 - 2y + 8 = 0$. 2 marks

b The circle is reflected in the y-axis.

What is the equation of the reflected circle? 1 mark

Question 19 (6 marks)

Xavier and Yen are participating in a charity bicycle ride between Canberra and Liverpool.

Xavier leaves Canberra and rides to Liverpool, a distance of 250 km, at an average speed of 20 km per hour. His distance from Liverpool is modelled by the equation $D = 250 - 20t$, where D is his distance from Liverpool and t is the time in hours he has been riding.

a On the grid below, draw the graph of this model and label it 'Xavier'. 1 mark

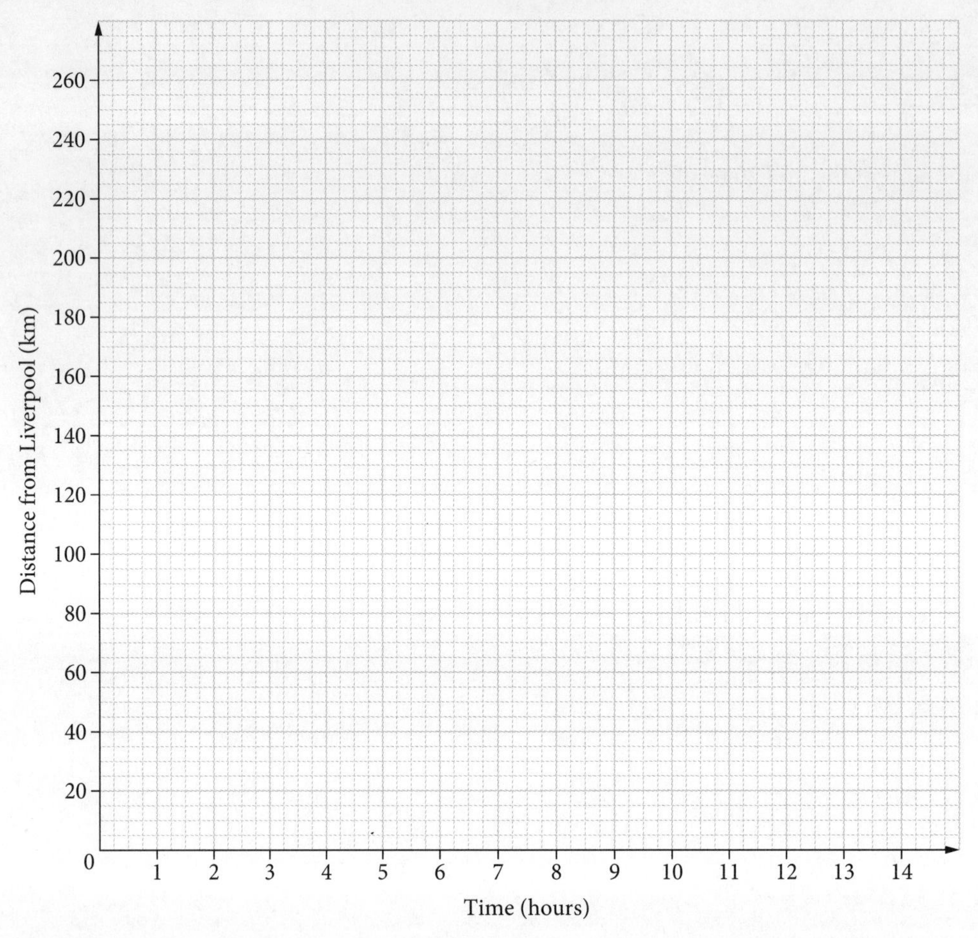

b Yen rides in the opposite direction and leaves from Berrima, a town located 90 km from Liverpool. He begins riding at the same time as Xavier and rides at an average speed of 12 km per hour towards Canberra.

By drawing a line on the grid above, or otherwise, find the value of t when Xavier and Yen pass each other. 2 marks

Question 19 continues on page 123

Question 19 (continued)

c Xavier and Yen are initially 160 km apart. Using the graphs drawn, or otherwise, find the value of t when Xavier and Yen are next 160 km apart. 1 mark

d Find the value of t when the riders have ridden a total of 264 km. 2 marks

End of Question 19

9780170459235

Question 20 (4 marks)

The gestation period of a human pregnancy is about 40 weeks. A study is conducted to investigate the correlation between gestation period (x weeks) and the birth weight (y grams) of the baby.

The table below shows the bivariate data for 12 babies, with the birth weight measured correct to the nearest 5 grams. The birth weight for Baby 5 is missing.

Baby	1	2	3	4	5	6	7	8	9	10	11	12
x weeks	34.7	36.0	29.3	40.1	35.7	40.3	37.3	40.9	38.3	41.4	39.7	38.0
y grams	1895	2030	1440	2835	*	3260	2690	3285	2920	3655	3685	2680

a Calculate, correct to four decimal places, Pearson's correlation coefficient using the 11 pairs of available data. 1 mark

b Calculate, correct to one decimal place, the mean gestation period, $\overline{x}$. 1 mark

c A least-squares regression line is fitted to the available data. Its equation is

$$y = -3918.3 + bx,$$

where b is a constant. The regression line passes through the point $(\overline{x}, 2789)$.

Use these facts to find, correct to one decimal place, the value of b. 1 mark

d Use the equation of the regression line to predict, to the nearest gram, the birth weight of Baby 5, indicated by * in the table. 1 mark

Question 21 (3 marks)

The table below shows the future value of an annuity of $1 for a selection of interest rates per period and investment terms. The contributions are made at the end of each period.

Period	Interest Rate Per Period											
	1.0%	2.0%	3.0%	4.0%	5.0%	6.0%	7.0%	8.0%	9.0%	10.0%	11.0%	12.0%
1	1.0000	1.0000	1.0000	1.0000	1.0000	1.0000	1.0000	1.0000	1.0000	1.0000	1.0000	1.0000
2	2.0100	2.0200	2.0300	2.0400	2.0500	2.0600	2.0700	2.0800	2.0900	2.1000	2.1100	2.1200
3	3.0301	3.0604	3.0909	3.1216	3.1525	3.1836	3.2149	3.2464	3.2781	3.3100	3.3421	3.3744
4	4.0604	4.1216	4.1836	4.2465	4.3101	4.3746	4.4399	4.5061	4.5731	4.6410	4.7097	4.7793
5	5.1010	5.2040	5.3091	5.4163	5.5256	5.6371	5.7507	5.8666	5.9847	6.1051	6.2278	6.3528
6	6.1520	6.3081	6.4684	6.6330	6.8019	6.9753	7.1533	7.3359	7.5233	7.7156	7.9129	8.1152
7	7.2135	7.4343	7.6625	7.8983	8.1420	8.3938	8.6540	8.9228	9.2004	9.4872	9.7833	10.0890
8	8.2857	8.5830	8.8923	9.2142	9.5491	9.8975	10.2598	10.6366	11.0285	11.4359	11.8594	12.2997
9	9.3685	9.7546	10.1591	10.5828	11.0266	11.4913	11.9780	12.4876	13.0210	13.5795	14.1640	14.7757
10	10.4622	10.9497	11.4639	12.0061	12.5779	13.1808	13.8164	14.4866	15.1929	15.9374	16.7220	17.5487
11	11.5668	12.1687	12.8078	13.4864	14.2068	14.9716	15.7836	16.6455	17.5603	18.5312	19.5614	20.6546
12	12.6825	13.4121	14.1920	15.0258	15.9171	16.8699	17.8885	18.9771	20.1407	21.3843	22.7132	24.1331

a Tess invests $800 at the end of each quarter into an account that earns 8% per annum with interest compounded quarterly. She does this for 3 years.

Find the future value of Tess' investment. 1 mark

b Her brother, Jake, finds an account that will earn him 12% per annum. The interest will also be compounded quarterly. Jake is only prepared to contribute for 2 years.

Find the amount that he must contribute quarterly if he is to achieve the same future value as Tess. 2 marks

Question 22 (6 marks)

The probability density function for a continuous random variable, X, is defined by

$$f(x) = \begin{cases} k(x-1)^2 & \text{for } 1 \le x \le 3 \\ 0 & \text{for all other values of } x \end{cases}.$$

a Find the value of k. 2 marks

b Find the median of the distribution, correct to two decimal places. 2 marks

c Find $P(X < 2)$. 2 marks

Questions 11–22 are worth 46 marks in total (Section II halfway point)

Question 23 (5 marks)

The diagram shows Q is 24 km due east of P.

Point R is 16 km from P on a bearing of 030°.

Find the bearing of R from Q, correct to the nearest degree. 5 marks

9780170459235

Question 24 (5 marks)

The percentage results of a national Science competition are normally distributed.

The mean mark is 64% and the standard deviation is 8.

a If Wendi's mark corresponds to a negative z-score in the range $-1 < z < 0$, give a possible value for her mark. 1 mark

b Chun obtains a result of 76%. Show that his z-score is 1.5. 1 mark

c Zoe's z-score was 1. What was her mark? 1 mark

d When Chun was sent his results, he was informed that his mark was better than that of 93.3% of competitors.

The competition was completed by 85 000 students. The shaded area below represents the students whose marks were higher than Zoe's but lower than Chun's.

How many students are represented by the shaded area? 2 marks

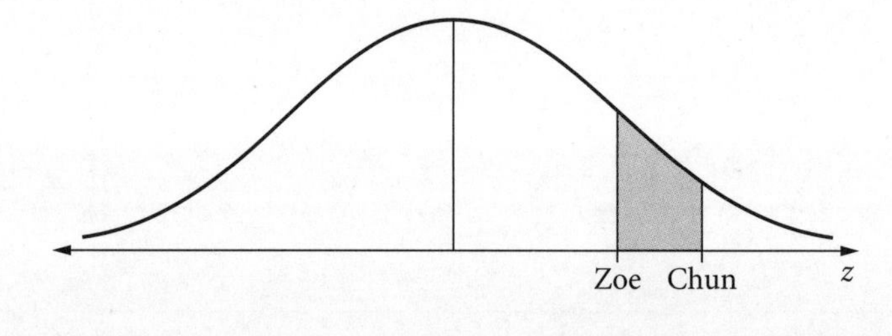

Question 25 (3 marks)

The diagram shows the graph of $y = 2\sqrt{x} - 1$. The area bounded by the graph, the x-axis, and the line $x = 4$ is shaded.

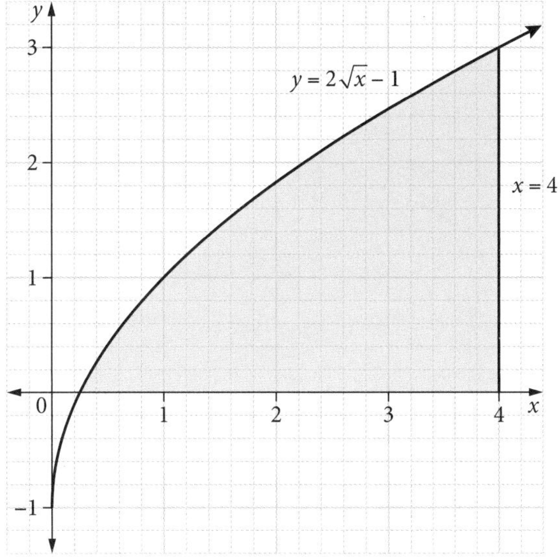

Find the exact area of the shaded region. 3 marks

Question 26 (6 marks)

The diagram below shows a right-angled triangular prism, *ABCDEF*, of fixed volume 540 cm^3.

The angles *ABC* and *EFD* are both 90°. The sides *AB* and *BC* are in the ratio 4 : 3.

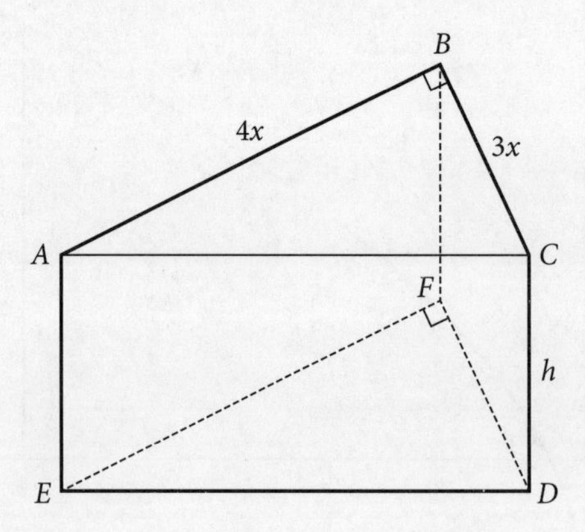

a Show that the surface area, *S*, of the prism is given by $S = 12x^2 + \dfrac{1080}{x}$. 3 marks

Question 26 continues on page 130

Question 26 (continued)

b Find the minimum surface area of the prism, to the nearest cm^2. 3 marks

End of Question 26

Question 27 (5 marks)

Adaline and Ben are twins. They both invest money in an account that earns 8% per annum which compounds annually.

a Adaline invests \$14 000 on 1 January. Find an expression, A_n, for the value of her investment after n years. 1 mark

b Ben decides instead to invest \$1800 each year with his first deposit also made on 1 January.

Show that the value of his investment after n years is given by 2 marks

$$B_n = \$24\,300 \times (1.08^n - 1).$$

c Show that the value of the twins' investments will be equal after approximately 11 years and 2 months. 2 marks

Question 28 (3 marks)

A discrete random variable, X, has the probability distribution shown below.

x	4	6	8	10	12
$P(x)$	0.3	0.22	0.17	m	0.1

a Find the value of m. 1 mark

b Find, correct to two decimal places, the value of the standard deviation, σ. 2 marks

Question 29 (4 marks)

The displacement of a particle moving along the x-axis is given by

$$x = e^t + 4e^{-t} - 3t - 5.$$

a Show that the particle is initially at the origin. 1 mark

b Find the exact time(s) when the particle comes to rest. 3 marks

Question 30 (5 marks)

The population of native parrots on a tropical island has been studied since 1950 for conservation purposes.

The number of parrots, P, can be modelled by the equation $P = 2500e^{-0.036t}$, where t is the number of years after 1950.

a What was the population of native parrots in 1970? 1 mark

b Show that the parrot population was declining at a rate of approximately 44 parrots per year in 1970. 2 marks

c When the parrot population reaches 250, they are listed as being in danger of extinction.

Use the model to predict the year this will occur. 2 marks

Question 31 (8 marks)

a Use the trapezoidal rule with 5 function values to approximate $\int_1^5 \ln x \, dx$. 2 marks

Give your answer correct to two decimal places.

Question 31 continues on page 134

Question 31 (continued)

b Find $\frac{d}{dx}(x\ln x)$. 2 marks

c Hence, or otherwise, find $\int_1^5 \ln x\, dx$. 3 marks

d Hence, find an approximation for $\ln 5$. 1 mark

END OF PAPER

WORKED SOLUTIONS

Section I (1 mark each)

Question 1

B $\sqrt{112} = \sqrt{16}\sqrt{7}$
$= 4\sqrt{7}$

Simple question.

Question 2

C As $3^4 = 81$ and $3^5 = 243$ for $3^x = 100$, x must be between 4 and 5.

Alternatively,

$\log_3 100 = \dfrac{\ln 100}{\ln 3}$ (by the change of base law)
≈ 4.2

The logarithm laws are important. They often appear in HSC exams.

Question 3

B The data points do not approximate a straight line with either a positive or negative gradient, so the only reasonable estimate for Pearson's correlation coefficient is –0.1.

New content. Common with Maths Standard 2.

Question 4

B As $f(3) = -6$, the graph of $y = f(x)$ passes through the point $(3, -6)$.

As $f'(3) = 0$, there is some kind of stationary point at $(3, -6)$.

As $f''(3) = 4$, the graph is concave up at $(3, -6)$, meaning the stationary point must be a local minimum.

Conceptual question. Band 5/6.

Question 5

A $\dfrac{\operatorname{cosec}\theta}{\cot\theta} = \operatorname{cosec}\theta \times \tan\theta$

$= \dfrac{1}{\sin\theta} \times \dfrac{\sin\theta}{\cos\theta}$

$= \dfrac{1}{\cos\theta}$

$= \sec\theta$

Know the trigonometric identities.

Question 6

C The test result is accurate for the 394 people with the virus who returned a positive test result.

The test result is also accurate for the 420 people without the virus who returned a negative test result.

$P(\text{test result is accurate}) = \dfrac{394 + 420}{930}$
≈ 0.8752
$\approx 88\%$

Common with Maths Standard 2.
Emphasis on literacy in so far as an understanding of what constitutes an 'accurate' result is required.

Question 7

B Consider the graph of the normal distribution below.

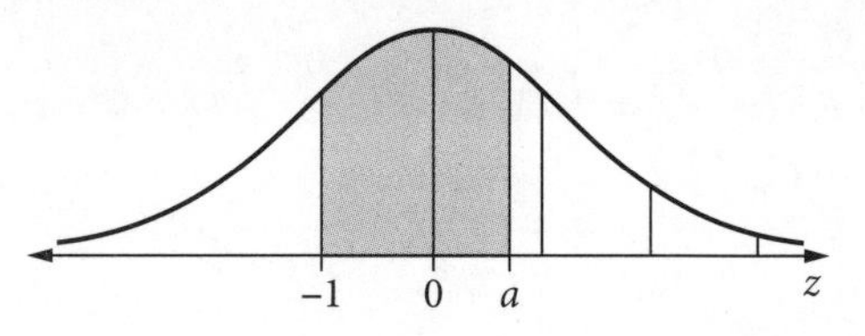

The shaded region represents all scores that have a z-score between –1 and a.

Since $P(z < a) = 0.7$, then $P(0 < z < a) = 0.2$ as 50% of scores have z-scores less than 0, by the symmetry of the normal distribution.

By the empirical rule on the reference sheet, $P(-1 < z < 1) = 0.68$, so $P(-1 < z < 0) = 0.34$, by symmetry.

So $P(-1 < z < a) = 0.34 + 0.2$
$= 0.54$

Common with Maths Standard 2, although this style of question is aimed at Maths Advanced students.
Quite simple to work out but lots of notation can make it hard to interpret.

Question 8

D

This is an application of the Fundamental Theorem of Calculus, which states that differentiation and integration are inverse operations.

Question 9

B $e^x(\cos x + 1) = 0$

So $e^x = 0$ OR $\cos x + 1 = 0$

No solution $\cos x = -1$

$x = \pi$

So the equation only has one solution in the given domain.

Remember that e^x is always positive.

Know your boundary angles, such as $\cos \pi$, not just 'All Stations To Central' for the quadrants.

Question 10

C The diagram shows a possible graph of $y = f(x)$.

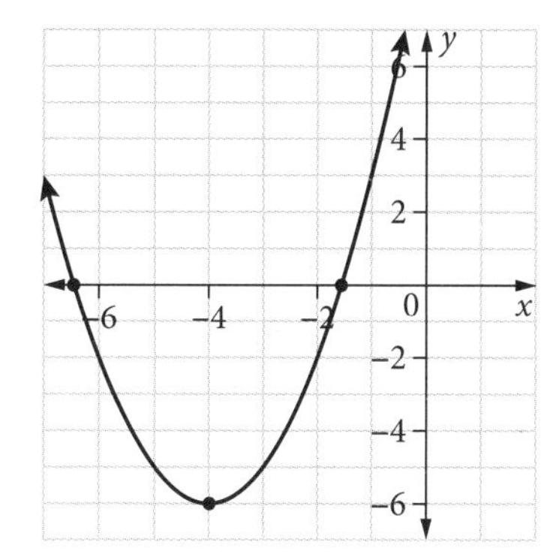

The graph of $y = -f(x)$ will be a reflection in the x-axis, producing a maximum turning point at $(-4, 6)$.

The horizontal dilation experienced by $y = -f(2x)$ will halve the distance of all points on $y = -f(x)$ from the y-axis.

Therefore, the maximum turning point at $(-4, 6)$ on the graph of $y = -f(x)$ will move to $(-2, 6)$ on the graph of $y = -f(2x)$.

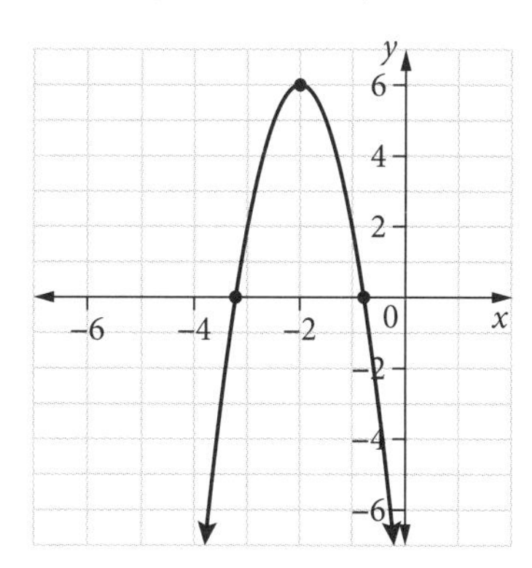

Transformations are new to the course.

Practise working with translations and reflections. Be extra careful with dilations. Many students find them difficult.

Section II (✓ = 1 mark)

Question 11 (4 marks)

a The coefficient of x shows the gradient of the line. ✓

b $c = 4$ (the constant term) ✓

c As the second line is perpendicular to $y = \frac{2}{3}x + 4$, its gradient must be $-\frac{3}{2}$. ✓

So apply the formula $y - y_1 = m(x - x_1)$.

$$y - 1 = -\frac{3}{2}(x + 2) \quad ✓$$
$$2y - 2 = -3x - 6$$
$$3x + 2y + 4 = 0$$

Straightforward questions in common with Maths Standard 2 (except part **c**) and Years 9–10 work.

Question 12 (2 marks)

$$\frac{d}{dx}\left(\frac{x^2}{x+1}\right) = \frac{(x+1)2x - x^2 \times 1}{(x+1)^2} \quad ✓$$
$$= \frac{2x^2 + 2x - x^2}{(x+1)^2}$$
$$= \frac{x^2 + 2x}{(x+1)^2} \quad ✓$$

Use of the quotient rule. Differentiate u first.

Remember to use the HSC exam reference sheet if needed.

Question 13 (2 marks)

$2x - 2y = 3$ [1]

$x + 4y = 9$ [2]

From [2],

$x = 9 - 4y$

Substitute into [1].

$$2(9 - 4y) - 2y = 3 \quad ✓$$
$$18 - 8y - 2y = 3$$
$$-10y = -15$$
$$y = 1\tfrac{1}{2}$$

So $x = 9 - 4\left(1\tfrac{1}{2}\right)$
$= 3$

So the solution is $x = 3, y = 1\tfrac{1}{2}$. ✓

Simultaneous equations were learned in Year 10. Check your answers carefully as you cannot afford to be making errors in questions such as these.

Question 14 (3 marks)

$$\int_0^{\frac{\pi}{3}} \cos 2x \, dx = \left[\frac{1}{2}\sin 2x\right]_0^{\frac{\pi}{3}} \quad ✓$$
$$= \left(\frac{1}{2}\sin\frac{2\pi}{3}\right) - 0 \quad ✓$$
$$= \frac{1}{2} \times \frac{\sqrt{3}}{2}$$
$$= \frac{\sqrt{3}}{4} \quad ✓$$

Simple integration question. By the HSC exam, you should have done many such trigonometric integrations.

Remember to use the HSC exam reference sheet if needed. Show the result of both substitutions, even if one of them is 0.

Question 15 (5 marks)

$y = x^3 - 4x^2$

So $y' = 3x^2 - 8x$.

Stationary points at $y' = 0$.

So $0 = 3x^2 - 8x$
$0 = x(3x - 8)$

So $x = 0$ or $x = \frac{8}{3}$.

When $x = 0, y = 0$.

When $x = \frac{8}{3}, y = \left(\frac{8}{3}\right)^3 - 4\left(\frac{8}{3}\right)^2$
$$= \frac{512}{27} - \frac{256}{9}$$
$$= -\frac{256}{27} \quad ✓$$

Determine the nature using y''.

Tell the marker what you are finding.

$y'' = 6x - 8$

When $x = 0, y'' = -8$
< 0

There is a maximum turning point at $(0, 0)$.

When $x = \frac{8}{3}, y'' = 8$
> 0

There is a minimum turning point at $\left(\frac{8}{3}, -\frac{256}{27}\right)$. ✓

Point of inflection at $y'' = 0$.

$$y'' = 6x - 8$$
$$0 = 6x - 8$$
$$6x = 8$$

So $x = \frac{4}{3}$.

When $x = \frac{4}{3}$, $y = \left(\frac{4}{3}\right)^3 - 4\left(\frac{4}{3}\right)^2$

$$= \frac{64}{27} - \frac{64}{9}$$
$$= -\frac{128}{27}$$

There is a possible point of inflection at $\left(\frac{4}{3}, -\frac{128}{27}\right)$.

x	1	$\frac{4}{3}$	2
y''	−2	0	4

Concavity changes and so $\left(\frac{4}{3}, -\frac{128}{27}\right)$ is confirmed as a point of inflection. ✓

Intercepts:

We have already established that $y = 0$ is the y-intercept.

Curve cuts x-axis when $y = 0$.

$$y = x^3 - 4x^2$$
$$0 = x^3 - 4x^2$$
$$0 = x^2(x - 4)$$

So $x = 0$ and 4. ✓

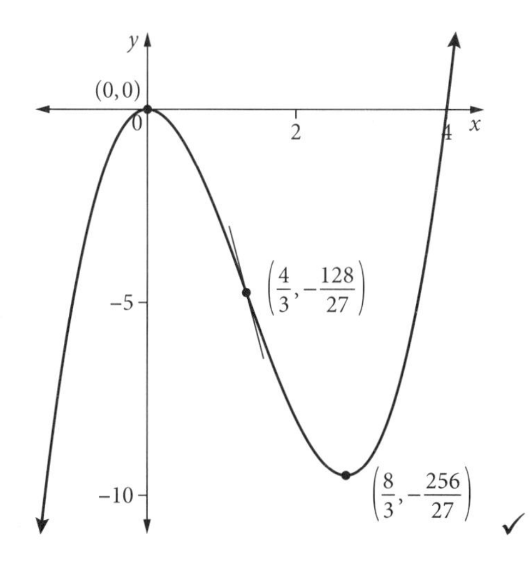

Typical 5-mark curve sketching question. Although one of the stationary points has a fractional x-value, this should not be an issue to senior students. Take care with your substitutions. Always check for a change in concavity before declaring a point of inflection. Always put numerical values in tables. Show precise details when drawing your graphs. Always take a ruler into any assessment task or major examination.

Question 16 (5 marks)

a

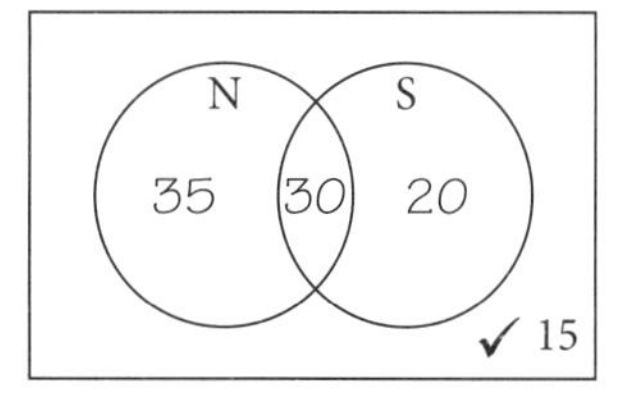

15 did not subscribe.
That leaves 100 − 15 = 85 who subscribe.

65 + 50 = 115 > 85
which means 115 − 85 = 30 have both.

Netflex only = 65 − 30 = 35
Span only = 50 − 30 = 20

With Venn diagrams, you need to work backwards.

b $P(\text{N} \cup \text{S}) = \frac{30}{100} = \frac{3}{10}$ ✓

c $P(\text{S} \mid \text{subscriber}) = \frac{50}{85} = \frac{10}{17}$ ✓

Conditional probability and Venn diagrams are new to the course. Practise using both.

d $P(\text{exactly one subscriber})$

$$= P(\text{S}, \tilde{\text{S}}) + P(\tilde{\text{S}}, \text{S})$$
$$= \left(\frac{85}{100} \times \frac{15}{99}\right) + \left(\frac{15}{100} \times \frac{85}{99}\right) \quad \checkmark \text{ per bracket}$$
$$= \frac{17}{66}$$

Straightforward question, but still has the potential to be done poorly.

Draw a tree diagram if needed.

Question 17 (3 marks)

a $T_{15} = a + 14d = 57$

$$8 + 14d = 57$$
$$14d = 49$$
$$d = 3.5 \checkmark$$

b $S_n = \frac{n}{2}[2a + (n-1)d]$

$$S_{20} = \frac{20}{2}[2(8) + (20-1)3.5] \checkmark$$
$$= 10 \times 82.5$$
$$= 825 \checkmark$$

Parts **a** and **b** are straightforward questions.
The formulas are on the HSC exam reference sheet.

Question 18 (3 marks)

a $x^2 - 8x + y^2 - 2y + 8 = 0$

$x^2 - 8x + 16 + y^2 - 2y + 1 = -8 + 16 + 1$

$(x - 4)^2 + (y - 1)^2 = 9$

Centre is $(4, 1)$ ✓ and radius is 3. ✓

b $(x + 4)^2 + (y - 1)^2 = 9$ ✓

Similar to Question 24 of the 2020 HSC exam. Questions involving completing the square to find the centre and radius of a circle are often asked on exams. Common errors include giving a centre of $(-4, -1)$ and a radius of 9.

Question 19 (6 marks)

a

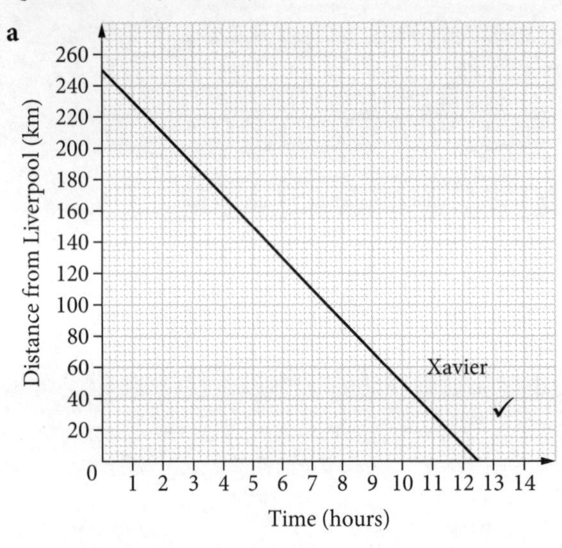

A line with y-intercept 250 and gradient -20, so it goes through $(10, 50)$.

Straightforward question.
Common content with Maths Standard 2.

b

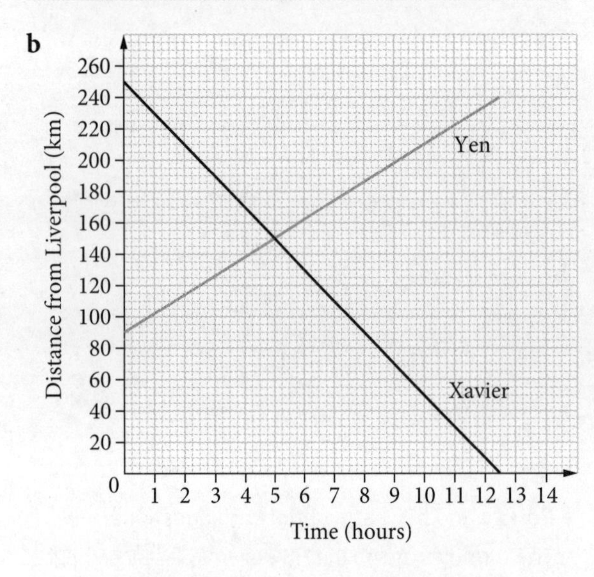

Yen's graph is a line with y-intercept 90 and gradient 12, so it goes through $(10, 210)$.

The point of intersection occurs at $t = 5$, so Xavier and Yen pass each other after 5 hours.

Alternatively, since Yen's journey is modelled by $D = 90 + 12t$, solve for t:

$250 - 20t = 90 + 12t$ ✓

$160 = 32t$

$t = 5$ ✓

Intersection of lines is common content with Maths Standard 2.
Expect to be examined on this topic.

c From the graph, $t = 10$ (when Xavier is 50 km and Yen is 210 km from Liverpool).

OR $90 + 12t - (250 - 20t) = 160$

$90 + 12t - 250 + 20t = 160$

$-160 + 32t = 160$

$32t = 320$

$t = 10$ hours ✓

Take care with the algebraic approach. Since D represents the distance from Liverpool, subtract Xavier from Yen and not the other way around, which will yield $t = 10$. Try it.

d The distances ridden by Xavier and Yen are $20t$ and $12t$ respectively.

$20t + 12t = 264$ ✓

$32t = 264$

$t = 8\frac{1}{4}$ hours ✓

Quite a complex problem. Expect that the last part of a question may be a little different.
Think before rushing in.

Question 20 (4 marks)

a $r \approx 0.9174$ ✓

b $\bar{x} \approx 37.8$ ✓

c $y = -3918.3 + bx$

Substitute $(37.8, 2789)$

$2789 = -3918.3 + b(37.8)$

So $b \approx 177.4$ ✓

d $y = -3918.3 + 177.4x$

When $x = 35.7$, $y = -3918.3 + 177.4(35.7)$

≈ 2415 grams ✓

Correlation and linear regression. All of Question 20 is common with Maths Standard 2.
Questions can sometimes be quite 'wordy'. Note that in part **c**, the statistics function of the calculator is *not* used for the least-squares regression line; algebraic methods are used instead.

Question 21 (3 marks)

a As interest is compound quarterly, $r = 2\%$ and $n = 12$.

Future value $= 800 \times 13.4121$
$= \$10\,729.68$ ✓

Straightforward question. Common with Maths Standard 2.
Always check the frequency of compounding.

b For Jake, $r = 3\%$ and $n = 8$. Let the amount of his quarterly contribution be $\$M$.

Then,

$10\,729.68 = M \times 8.8923$ ✓
$M = \$1206.63$ ✓

Straightforward question. Common with Maths Standard 2. Future and Present Value tables are a recent addition to the Advanced course.

Question 22 (6 marks)

a The area under the PDF must be 1.

$$\int_1^3 k(x-1)^2 dx = 1$$

$$k\int_1^3 (x-1)^2\, dx = 1 \quad ✓$$

$$k\left[\frac{(x-1)^3}{3}\right]_1^3 = 1$$

$$k\left(\frac{8}{3} - 0\right) = 1$$

$$k = \frac{3}{8} \quad ✓$$

b $\text{CDF} = \int_1^x \frac{3}{8}(t-1)^2\, dt$

$$= \frac{3}{8}\left[\frac{(t-1)^3}{3}\right]_1^x$$

$$= \left(\frac{(x-1)^3}{3}\right) - (0)$$

$$= \frac{(x-1)^3}{8} \quad ✓$$

To find the median, solve $\text{CDF} = \frac{1}{2}$.

So $\frac{(x-1)^3}{8} = \frac{1}{2}$

$$(x-1)^3 = 4$$
$$x - 1 = \sqrt[3]{4}$$
$$x = \sqrt[3]{4} + 1$$
$$\approx 2.59 \quad ✓$$

c $P(X < 2) = \int_1^2 \frac{3}{8}(x-1)^2\, dx$ ✓

$$= \left[\frac{(x-1)^3}{8}\right]_1^2$$

$$= \frac{1}{8} - 0$$

$$= \frac{1}{8} \quad ✓$$

PDFs and CDFs are new to the course.
Expect that they will be examined regularly.

Question 23 (5 marks)

$\angle RPQ = 90° - 30° = 60°$ ✓

By cosine rule,

$RQ^2 = 16^2 + 24^2 - 2 \times 16 \times 24 \times \cos 60°$ ✓
$= 448$

So $RQ = \sqrt{448}$
$= 8\sqrt{7}$ ✓

Again, by the cosine rule,

$$\cos \angle RQP = \frac{p^2 + r^2 - q^2}{2pr}$$

$$= \frac{(8\sqrt{7})^2 + 24^2 - 16^2}{2 \times 8\sqrt{7} \times 24}$$

$$= 0.7559\ldots$$

$$\angle RQP \approx 41° \quad ✓$$

So, the bearing of R from Q is approximately $270° + 41° = 311°$. ✓

A 5-mark question with no parts/clues. Common with Maths Standard 2.
Watch for questions for which no scaffolding has been provided. You will be expected to negotiate your own way through the solution. Spend a minute planning your answer rather than writing straight away.
Many students have trouble with bearings. Examiners are aware of this and will often ask it. Practise!

Question 24 (5 marks)

a $64 - 8 = 56$, so a possible mark is 59%, or any number between 56% and 64%. ✓

b $z = \frac{x - \mu}{\sigma}$

$$= \frac{76 - 64}{8} \quad ✓$$

$= 1.5$, as required

c Making x the subject of the z-score formula:

$$x = \mu + z\sigma$$
$$= 64 + 1(8) \checkmark$$
$$= 72$$

All of Question 24 is common with Maths Standard 2.

d With a z-score of 1, Zoe's result will be better than $50\% + \frac{68\%}{2} = 84\%$ of scores.

As Chun's result is better than 93.3% of scores, the percentage between Zoe and Chun is $93.3\% - 84\% = 9.3\%$. ✓

Students represented by the shaded region
$= 9.3\% \times 85\,000$
$= 7905$ ✓

A challenging question with some extra interpretation required.

Not mathematically difficult but it's easy to get lost in the explanation given in the question.

Question 25 (3 marks)

x-intercept when $y = 0$

$$0 = 2\sqrt{x} - 1$$
$$1 = 2\sqrt{x}$$
$$\sqrt{x} = \frac{1}{2}$$
$$x = \left(\frac{1}{2}\right)^2 = \frac{1}{4} \checkmark$$

$$\text{Area} = \int_{\frac{1}{4}}^{4} 2\sqrt{x} - 1\,dx$$
$$= \left[\frac{2}{\frac{3}{2}}x^{\frac{3}{2}} - x\right]_{\frac{1}{4}}^{4} \checkmark$$
$$= \left[\frac{4}{3}x^{\frac{3}{2}} - x\right]_{\frac{1}{4}}^{4}$$
$$= \left[\frac{4}{3}(4)^{\frac{3}{2}} - 4\right] - \left[\frac{4}{3}\left(\frac{1}{4}\right)^{\frac{3}{2}} - \frac{1}{4}\right]$$
$$= \left[\frac{4}{3}(8) - 4\right] - \left[\frac{4}{3}\left(\frac{1}{8}\right) - \frac{1}{4}\right]$$
$$= \frac{27}{4} \text{ units}^2 \checkmark$$

Area of integration question that is aimed at Band 5/6 students. There is no deep conceptual thinking involved, but the algebraic manipulation is a little more sophisticated. Solution requires use of fractional indices and substitutions.

Question 26 (6 marks)

a $\text{Volume} = \frac{1}{2}(3x)(4x)h$

$$540 = \frac{1}{2}(3x)(4x)h$$
$$540 = 6x^2h$$
$$90 = x^2h$$
$$h = \frac{90}{x^2} \checkmark$$

By Pythagoras' theorem, $AC^2 = (3x)^2 + (4x)^2$

so $AC = \sqrt{(3x)^2 + (4x)^2}$
$= 5x$ ✓

$$S = 2\left(\frac{1}{2} \times 3x \times 4x\right) + 3xh + 4xh + 5xh$$
$$= 12x^2 + 12xh$$

But $h = \frac{90}{x^2}$

so $S = 12x^2 + 12x\left(\frac{90}{x^2}\right)$ ✓

$= 12x^2 + \frac{1080}{x}$, as required.

Targeted at Band 6.

Optimisation is a difficult part of the course and is almost always examined.

The high level of algebraic skill needed can also prove challenging. Work on your algebra.

b $S = 12x^2 + \frac{1080}{x}$

$$= 12x^2 + 1080x^{-1}$$

So $S' = 24x - 1080x^{-2}$ ✓

$$= 24x - \frac{1080}{x^2}$$

S is maximised when $S' = 0$.

So $0 = 24x - \frac{1080}{x^2}$

$$= 24x^3 - 1080$$
$$= x^3 - 45$$
$$x^3 = 45$$
$$x = \sqrt[3]{45}$$

Now,

$$S'' = 24 + 2160x^{-3}$$
$$= 24 + \frac{2160}{x^3}$$

When $x = \sqrt[3]{45}$

$$S'' = 24 + \frac{2160}{45}$$
$$= 72$$
$$> 0$$

So S is minimised when $x = \sqrt[3]{45}$. ✓

$$\text{Minimum surface area} = 12\left(\sqrt[3]{45}\right)^2 + \frac{1080}{\sqrt[3]{45}}$$
$$\approx 455\text{ cm}^2 \checkmark$$

Most students should at least be able to differentiate the expression for S.
Write something even if you don't think you can finish the question.
Remember: 'Blank answers score 0'.

Question 27 (5 marks)

a $A_n = 14000(1 + 0.08)^n$
$A_n = 14000(1.08)^n$ ✓

Straightforward question, but applied in a general context and not a specific one.
The question asks for an algebraic expression and not a numerical answer, which confuses some students.
Always expect a Financial Maths question, especially since much of the content that was previously only found in Maths Standard 2 is now examinable in Maths Advanced as well.

b $B_1 = 1800(1.08)$

$$B_2 = (1800(1.08) + 1800)1.08$$
$$= 1800(1.08)^2 + 1800(1.08)$$
$$= 1800(1.08^2 + 1.08) \checkmark$$

$$B_3 = (1800(1.08)^2 + 1800(1.08) + 1800)1.08$$
$$= 1800(1.08)^3 + 1800(1.08)^2 + 1800(1.08)$$
$$= 1800(1.08^3 + 1.08^2 + 1.08)$$

Generalising this pattern,

$$B_n = 1800(1.08 + 1.08^2 + 1.08^3 + \cdots + 1.08^n)$$
$$= 1800\left[\frac{1.08(1.08^n - 1)}{1.08 - 1}\right] \quad \text{(by summing the geometric series)}$$
$$= 1800[13.5(1.08^n - 1)]$$
$$= 24300(1.08^n - 1), \text{ as required } \checkmark$$

Familiar process but applied in a general sense where an expression is required.
Is B_3 required so that you have been deemed to establish the pattern? Possibly not, but better to be safe.
Notice that there are 3 terms written in the series for B_n before the ellipsis ($\cdots$). *Never* stop at 2. You need 3 to establish a geometric series.

c $14000(1.08)^n = 24300(1.08^n - 1)$ ✓

$$14000(1.08)^n = 24300(1.08)^n - 24300$$
$$10300(1.08)^n = 24300$$
$$1.08^n = \frac{103}{243}$$
$$\text{So } n = \frac{\ln\left(\frac{103}{243}\right)}{\ln(1.08)}$$
$$\approx 11.1528 \text{ years}$$
$$\approx 11 \text{ years } 1.833\ldots \text{ months}$$
$$\approx 11 \text{ years } 2 \text{ months},$$

as required ✓

Higher-level question involving an exponential equation.
Well-developed algebraic skills are required.

Question 28 (3 marks)

a $0.3 + 0.22 + 0.17 + m + 0.1 = 1$

So $m = 0.21$. ✓

Typical discrete probability question.

b $\mu = 4(0.3) + 6(0.22) + 8(0.17) + 10(0.21) + 12(0.1)$
$= 7.18$ ✓

$$\text{Var}(X) = E(X^2) - \mu^2$$
$$= 16(0.3) + 36(0.22) + 64(0.17) + 100(0.21) + 144(0.1) - 7.18^2$$
$$= 7.4476$$

$$\sigma = \sqrt{\text{Var}(X)} \approx 2.73 \checkmark$$

New content. Practise these processes.

Question 29 (4 marks)

a $x = e^t + 4e^{-t} - 3t - 5$

When $t = 0$,

$$x = e^0 + 4e^0 - 3(0) - 5 \checkmark$$
$$= 1 + 4 - 0 - 5$$
$$= 0$$

So the particle is initially at the origin, as required.

Straightforward substitution.
Remember, $e^0 = 1$, not 0.

WORKED SOLUTIONS

b $x = e^t + 4e^{-t} - 3t - 5$

So $v = e^t - 4e^{-t} - 3$. ✓

When $v = 0$,

$$\begin{aligned} 0 &= e^t - 4e^{-t} - 3 \\ &= e^t - \frac{4}{e^t} - 3 \\ &= e^{2t} - 3e^{-t} - 4 \\ &= (e^t - 4)(e^t + 1) \end{aligned}$$

So $e^t = 4$ OR $e^t = -1$. ✓

$t = \ln 4$ No solution

So the particle is at rest after $\ln 4$ seconds only. ✓

Leave answer as ln 4 because an exact answer was required. Band 6-style question.

Strong algebra skills required.

Question 30 (5 marks)

a $P = 2500e^{-0.036t}$

When $t = 20$,

$$\begin{aligned} P &= 2500e^{-0.036(20)} \\ &\approx 1217 \text{ parrots} \end{aligned}$$ ✓

Round to nearest whole number.

b $P = 2500e^{-0.036t}$

So $\frac{dP}{dt} = -90e^{-0.036t}$ ✓

When $t = 20$,

$$\begin{aligned} \frac{dP}{dt} &= -90e^{-0.036(20)} \\ &= -43.8077\ldots \\ &\approx -44 \text{ parrots/year} \end{aligned}$$ ✓

Make sure you can differentiate all function types.

c $P = 2500e^{-0.036t} < 250$ ✓

$$\begin{aligned} e^{-0.036t} &< 0.1 \\ -0.036t &< \ln(0.1) \\ t &> 63.96 \end{aligned}$$

$t = 64$ corresponds to the year 2014, since we are measuring from 1950.

So in 2014, the parrots will be in danger of extinction. ✓

Band 6-level question. Strong algebra needed.

Don't forget to answer the question with the correct year; don't stop at $t = 64$. Note also the reversal of the inequality sign because we divided by ln (0.1), which is negative.

Question 31 (8 marks)

a

x	1	2	3	4	5
$f(x)$	0	$\ln 2$	$\ln 3$	$\ln 4$	$\ln 5$

$$\begin{aligned} \int_1^5 \ln x\, dx &\approx \frac{1}{2}[0 + 2(\ln 2 + \ln 3 + \ln 4) + \ln 5] \\ &\approx 3.98 \end{aligned}$$ ✓ ✓

It is rare for the trapezoidal rule to not appear in the HSC exam. Be aware that it may be in a practical context in a shared question with Maths Standard 2.

b $$\begin{aligned} \frac{d}{dx}(x \ln x) &= x\left(\frac{1}{x}\right) + \ln x(1) \\ &= 1 + \ln x \end{aligned}$$ ✓ ✓

Just because it's the last question, don't assume it will all be too hard for you.

c From part **b**,

$$\begin{aligned} \int_1^5 (1 + \ln x)\, dx &= [x \ln x]_1^5 \quad ✓ \\ \int_1^5 1\, dx + \int_1^5 \ln x\, dx &= [x \ln x]_1^5 \\ \int_1^5 (\ln x)\, dx &= [x \ln x]_1^5 - \int_1^5 1\, dx \quad ✓ \\ \int_1^5 (\ln x)\, dx &= [x \ln x]_1^5 - [x]_1^5 \\ &= (5\ln 5 - 0) - (5 - 1) \\ &= 5\ln 5 - 4 \quad ✓ \end{aligned}$$

Band 6-level question.

The inverse relationship between differentiation and integration is a common question style. See Question 13(c) from the 2019 HSC exam or Question 12(d) from 2016.

d From parts **a** and **c**,

$$\begin{aligned} 5\ln 5 - 4 &\approx 3.98 \\ 5\ln 5 &\approx 7.98 \end{aligned}$$

So $\ln 5 \approx 1.596$. ✓

The answers to previous parts of a question are often used in later parts. Examiners have reasons for asking certain questions. It's never random!

Mathematics Advanced

PRACTICE HSC EXAM 2

General instructions

- Reading time: 10 minutes
- Working time: 3 hours
- A reference sheet is provided on page 195 at the back of this book
- For questions in Section II, show relevant mathematical reasoning and/or calculations

Total marks: 100

Section I – 10 questions, 10 marks

- Attempt Questions 1–10
- Allow about 15 minutes for this section

Section II – 21 questions, 90 marks

- Attempt Questions 11–31
- Allow about 2 hours and 45 minutes for this section

9780170459235

Section I

10 marks
Attempt Questions 1–10
Allow about 15 minutes for this section

Circle the correct answer.

Question 1

A line passes through the points $A(4,-2)$ and $B(1,3)$. What is its gradient?

A $-\frac{5}{3}$

B $-\frac{3}{5}$

C $\frac{3}{5}$

D $\frac{5}{3}$

Question 2

What is the value of p if $\frac{a^4\sqrt{a}}{a^{-1}} = a^p$?

A 2

B 3

C $\frac{7}{2}$

D $\frac{11}{2}$

Question 3

Which of the following is an example of a discrete random variable?

A The time needed to swim 200 metres.

B The number of visits made to a cinema in a year.

C The height of a Year 12 student.

D The mass of a parcel.

Question 4

If $y = \cos(x^2)$, what is $\frac{dy}{dx}$?

A $2x\cos(x^2)$

B $-2x\cos(x^2)$

C $2x\sin(x^2)$

D $-2x\sin(x^2)$

Question 5

Frank deposits \$200 at the end of each month into an annuity earning 6% per annum with interest compounded monthly. He does this for 3 years.

The table shows the future value of an annuity of \$1 for a selection of interest rates per period and investment terms.

	Interest rate per period											
Period	**0.5%**	**1.0 %**	**1.5 %**	**2.0%**	**2.5%**	**3.0%**	**3.5 %**	**4.0 %**	**4.5 %**	**5.0 %**	**5.5 %**	**6.0 %**
3	3.0150	3.0301	3.0452	3.0604	3.0756	3.0909	3.1062	3.1216	3.1370	3.1525	3.1680	3.1836
6	6.0755	6.1520	6.2296	6.3081	6.3877	6.4684	6.5502	6.6330	6.7169	6.8019	6.8881	6.9753
9	9.1821	9.3685	9.5593	9.7546	9.9545	10.1591	10.3685	10.5828	10.8021	11.0266	11.2563	11.4913
12	12.3356	12.6825	13.0412	13.4121	13.7956	14.1920	14.6020	15.0258	15.4640	15.9171	16.3856	16.8699
15	15.5365	16.0969	16.6821	17.2934	17.9319	18.5989	19.2957	20.0236	20.7841	21.5786	22.4087	23. 2760
18	18.7858	19.6147	20.4894	21.4123	22.3863	23.4144	24.4997	25.6454	26.8551	28.1324	29.4812	30.9057
21	22.0840	23.2392	24.4705	25.7833	27.1833	28.6765	30.2695	31.9692	33.7831	35.7193	37.7861	39.9927
24	25.4320	26.9735	28.6335	30.4219	32.3490	34.4265	36.6665	39.0826	41.6892	44.5020	47.5380	50 .8156
27	28.8304	30.8209	32.9867	35.3443	37.9120	40.7096	43.7591	47.0842	50.7113	54.6691	58.9891	63.7058
30	32.2800	34.7849	37.5387	40.5681	43.9027	47.5754	51.6227	56.0849	61.0071	66.4388	72.4355	79.0582
33	35.7817	38.8690	42.2986	46.1116	50.3540	55.0778	60.3412	66.2095	72.7562	80.0638	88.2248	97.3432
36	39.3361	43.0769	47.2760	51.9944	57.3014	63.2759	70.0076	77.5983	86.1640	95.8363	106.7652	119.1209
39	42.9441	47.4123	52.4807	58.2372	64.7830	72.2342	80.7249	90.4091	101.4644	114.0950	128.5361	145.0585
42	46.6065	51.8790	57.9231	64.8622	72.8398	82.0232	92.6074	104.8196	118.9248	135. 2318	154 .1005	175.9505

What is the interest earned on Frank's investment?

A \$36.72

B \$636.72

C \$667.22

D \$7867.22

Question 6

The graph shows the displacement, x, of a particle moving along a straight line as a function of time, t.

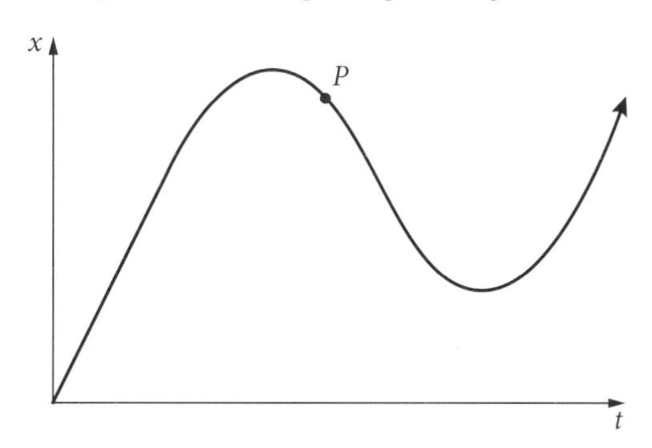

Which statement describes the motion of the particle at point P?

A The particle is moving to the right with increasing speed.

B The particle is moving to the left with increasing speed.

C The particle is moving to the right with decreasing speed.

D The particle is moving to the left with decreasing speed.

Question 7

Given that $2\log a = \log b - \log 3$, where $a > 0$ and $b > 0$, which is the correct expression for a?

A $a = \dfrac{b-3}{2}$

B $a = \sqrt{b-3}$

C $a = \sqrt{\dfrac{b}{3}}$

D $a = \dfrac{b}{6}$

Question 8

The probability that John is at school on a school day is $\frac{19}{20}$, whereas the probability that Lucas is at school on a school day is $\frac{14}{15}$.

What is the probability that both boys are absent (not at school) on a particular day?

A $\dfrac{1}{300}$

B $\dfrac{19}{300}$

C $\dfrac{71}{150}$

D $\dfrac{133}{150}$

Question 9

If the numbers 2, x, 10 are the first 3 terms of a geometric series, which of the following is a possible value of x?

A $-\sqrt{20}$

B 5

C 6

D $2\sqrt{10}$

Question 10

For what range of values of θ is $\tan\theta + \cos\theta < 0$?

A $\left(0, \dfrac{\pi}{2}\right)$

B $\left(\dfrac{\pi}{2}, \pi\right)$

C $\left(\pi, \dfrac{3\pi}{2}\right)$

D $\left(\dfrac{3\pi}{2}, 2\pi\right)$

Section II

90 marks
Attempt Questions 11–31
Allow about 2 hours and 45 minutes for this section

- Answer the questions in the spaces provided. These spaces provide guidance for the expected length of response.
- Your responses should include relevant mathematical reasoning and/or calculations.

Question 11 (2 marks)

Factorise fully $4x - xy^2$. 2 marks

Question 12 (3 marks)

The solutions to the quadratic equation $x^2 - 4x + 1 = 0$ are $a \pm \sqrt{b}$. Find a and b. 3 marks

9780170459235

Question 13 (2 marks)

Find the equation of the tangent to the graph of $y = 2x^2 - 3$ at the point $(1, -1)$. 2 marks

Question 14 (2 marks)

The table below shows the blood types of Australians by percentage.

Type	A	B	O	AB
%	38	10	49	3

Three people are selected at random. Find the probability that:

a none of the three have blood type B. 1 mark

b at least one of them has blood type B. 1 mark

Question 15 (2 marks)

Find the exact value of $\sec\left(\frac{3\pi}{4}\right)$. 2 marks

Question 16 (4 marks)

Karen opens a flower shop. During her first month of business, she had sales totalling $9000.

She notices that her sales in each subsequent month are only 95% of the sales in the previous month.

a Find the value of her sales in the 3rd month. 1 mark

b Show that Karen's total sales will never exceed $180 000. 1 mark

c What percentage, correct to one decimal place, of Karen's total sales are made in the first 4 months of trading? 2 marks

Question 17 (6 marks)

The circle in the diagram below has centre $N(-4, -4)$.

The line l is a tangent to the circle at P and has the equation $3x + 2y + 7 = 0$.

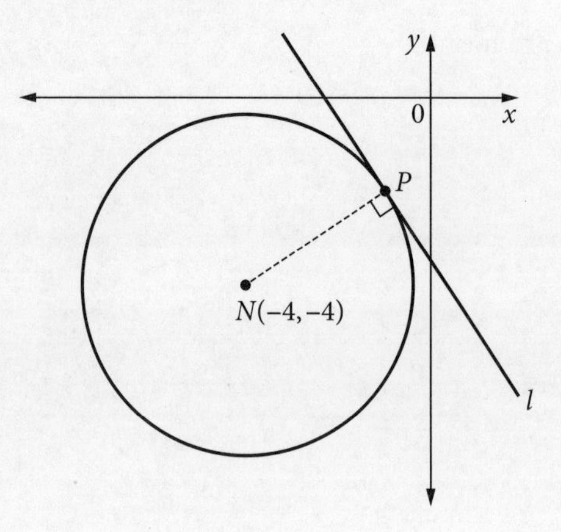

a Show that the equation of NP is $2x - 3y - 4 = 0$. 2 marks

b Find the coordinates of point P. 2 marks

Question 17 continues on page 153

Question 17 (continued)

c Find the exact radius of the circle. 1 mark

d Write the equation of the circle. 1 mark

Question 18 (5 marks)

The diagram below represents the course sailed in a yacht race.

The yachts begin at A and then proceed to round the buoys at B, C, D and E, in turn, before returning to the finish line at A.

Some of the angles and distances are shown.

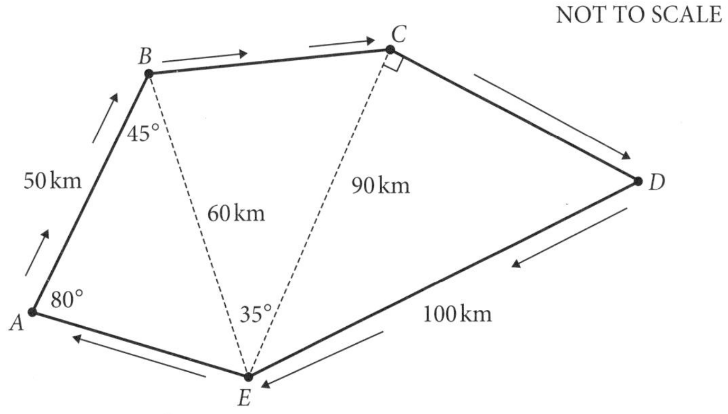

NOT TO SCALE

B
C
D
E
A
45°
80°
35°
50 km
60 km
90 km
100 km

Find the length of the course, correct to the nearest 10 km. 5 marks

Question 19 (3 marks)

Nicky wants to take out a loan to buy an investment property.

The table below shows the monthly repayments required for a variety of loan amounts in increments of $10 000 over varying terms.

	Term of loan in months				
Principal	**10 years**	**15 years**	**20 years**	**25 years**	**30 years**
$200 000	2101.81	1560.83	1297.91	1145.99	1049.33
$210 000	2206.90	1638.87	1362.81	1203. 29	1101.80
$220 000	2311.99	1716.91	1427.71	1260.59	1154.26
$230 000	2417.08	1794.95	1492.60	1317.89	1206.73
$240 000	2522.17	1872.99	1557.50	1375.19	1259.20
$250 000	2627.27	1951.04	1622.39	1432.49	1311.66
$260 000	2732.36	2029.08	1687.29	1489.79	1364.13
$270 000	2837.45	2107.12	1752.19	1547.09	1416.60
$280 000	2942.54	2185.16	1817.08	1604.39	1469.06
$290 000	3047.63	2263.20	1881.98	1661.69	1521.53
$300 000	3152.72	2341.24	1946.87	1718.99	1574.00
$310 000	3257.81	2419.28	2011.77	1776.29	1626.46
$320 000	3362.90	2497.33	2076.66	1833.59	1678.93
$330 000	3467.99	2575.37	2141.56	1890.89	1731.40
$340 000	3573.08	2653.41	2206.46	1948.19	1783.86
$350 000	3678.17	2731.45	2271.35	2005.49	1836.33
$360 000	3783.26	2809.49	2336.25	2062 .79	1888 .80
$370 000	3888.35	2887.53	2401.14	2120 .09	1941.26

a Nicky decides to borrow $300 000 over a 20-year period.

Find her monthly repayment. 1 mark

b How much interest will she pay over the term of the loan? 1 mark

c How much money would Nicky save if she reduced the term of the loan to 15 years? 1 mark

Question 20 (2 marks)

The curve below shows the rate of population growth, $\frac{dP}{dt}$, in a coastal town over a period of 10 weeks.

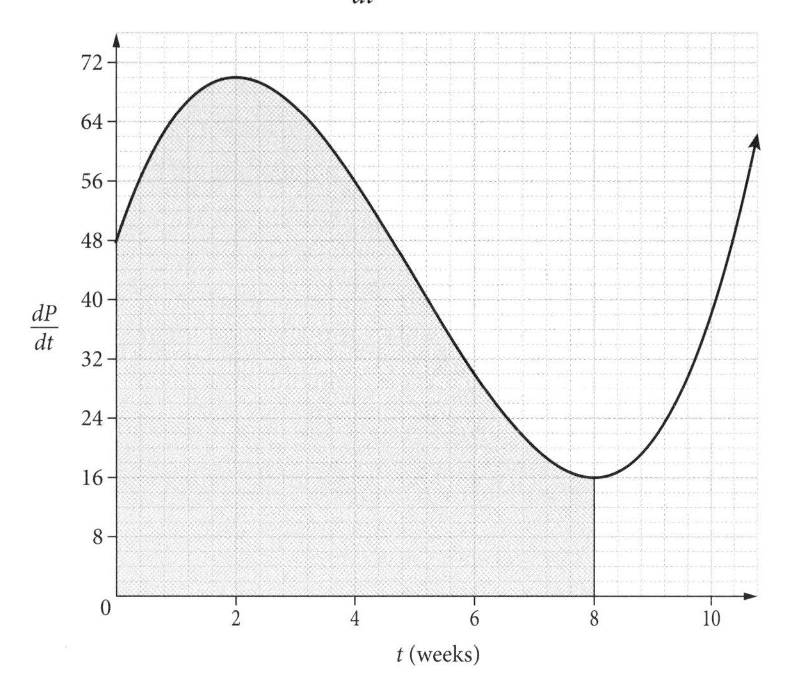

Using the trapezoidal rule with 5 function values, estimate the number of people who moved to the town during the 8-week period represented by the shaded region. 2 marks

Question 21 (5 marks)

A bag contains 9 red and 7 blue marbles. Two marbles are selected at random from the bag.

If the first marble is red, its colour is noted and it is replaced by a blue marble before the second marble is selected.

Similarly, if the first marble is blue, its colour is noted and it is replaced by a red marble before the second marble is selected.

By constructing a probability tree diagram, or otherwise, find the probability that:

a the first marble selected is blue. 1 mark

b the second marble is blue, given that the first is blue. 1 mark

c both marbles selected are red. 1 mark

d the marbles selected are different colours. 2 marks

9780170459235

Question 22 (2 marks)

Given that $f(x) = x^2 + 1$ and $g(x) = \sqrt{x - 3}$, sketch $y = f(g(x))$ over its natural domain. 2 marks

Question 23 (4 marks)

a Show that $\cos^2\theta\,(\sec^2\theta - 1) = \sin^2\theta$. 2 marks

b Hence, solve $4\cos^2\theta\,(\sec^2\theta - 1) = 1$ for θ in the domain $[0, 2\pi]$. 2 marks

Question 24 (6 marks)

Surinder is planning for his retirement and has been contributing to a superannuation fund.

His financial advisor informs him that he has two options.

Option A	Retire at the age of 66 and receive a defined pension of \$600 payable 25 times per year at equal time intervals.
Option B	Retire at the age of 70 and receive a defined pension of \$760 payable 25 times per year at equal time intervals.

Let n be Surinder's age in years and A be the total amount of pension received.

a If Surinder chooses to retire at 66, the total of his pension received can be modelled by

$$A = 15\,000(n - 66).$$

On the grid below, draw the graph of this model and label it 'Option A'. 2 marks

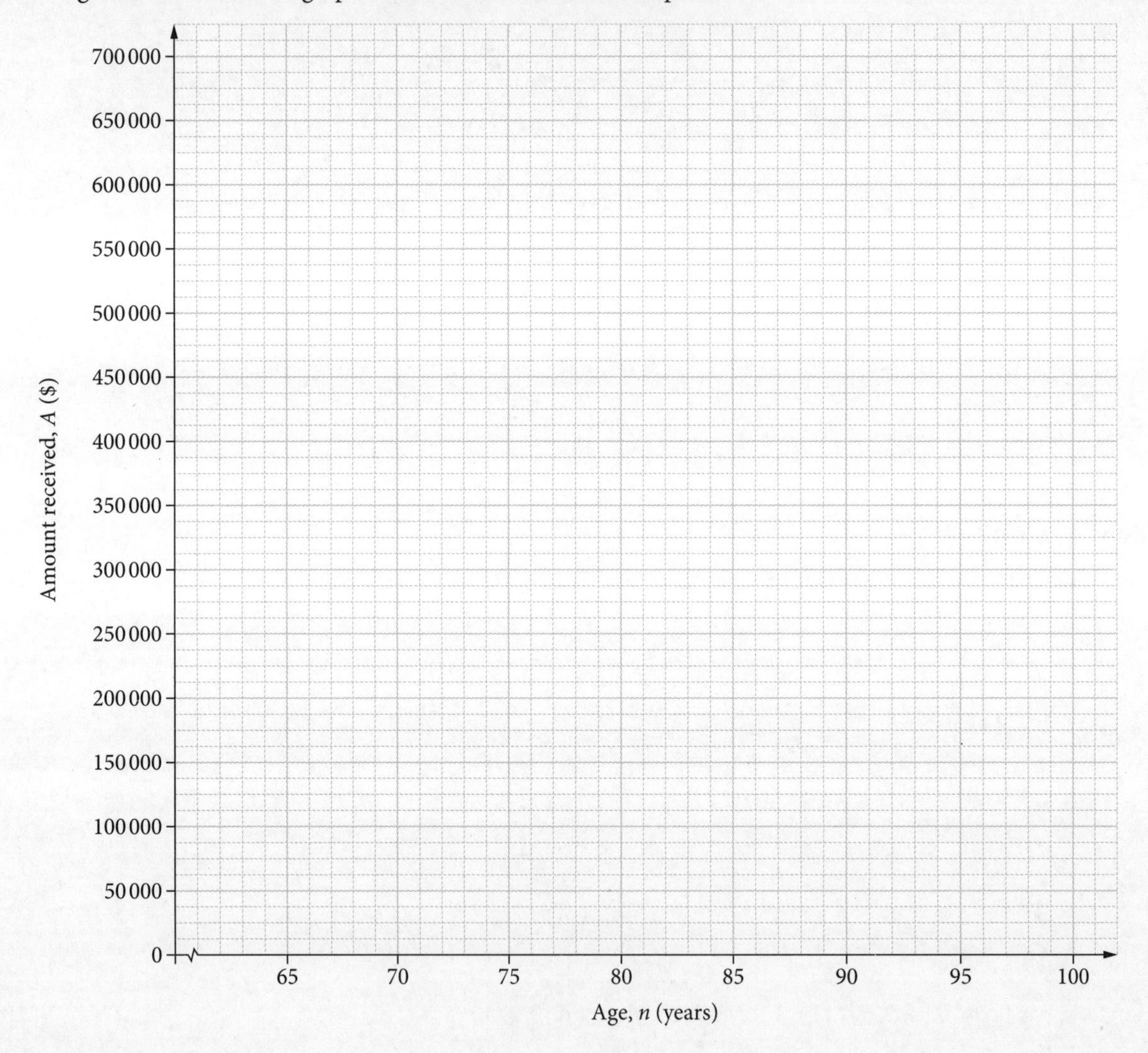

Question 24 continues on page 159

Question 24 (continued)

b Suppose Surinder chooses Option B.

i Find the total amount of pension he will receive by the time he is 80 years old. 1 mark

ii Write a rule connecting A and n that will accurately model his total pension received at age n. 1 mark

c By drawing the graph for Option B on the graph for Option A on the previous page, or otherwise, find how old Surinder will be when his total pension received will be the same for both Option A and Option B. 2 marks

Questions 11–24 are worth 48 marks in total (Section II halfway point)

Question 25 (5 marks)

The average temperature, T, in Melbourne on a day in April is modelled by the equation

$$T = 8\sin\left[\frac{\pi}{12}(t-4)\right] + 12,$$

where t is the time in hours after 6 am and $0 \le t \le 24$.

The graph of T appears below.

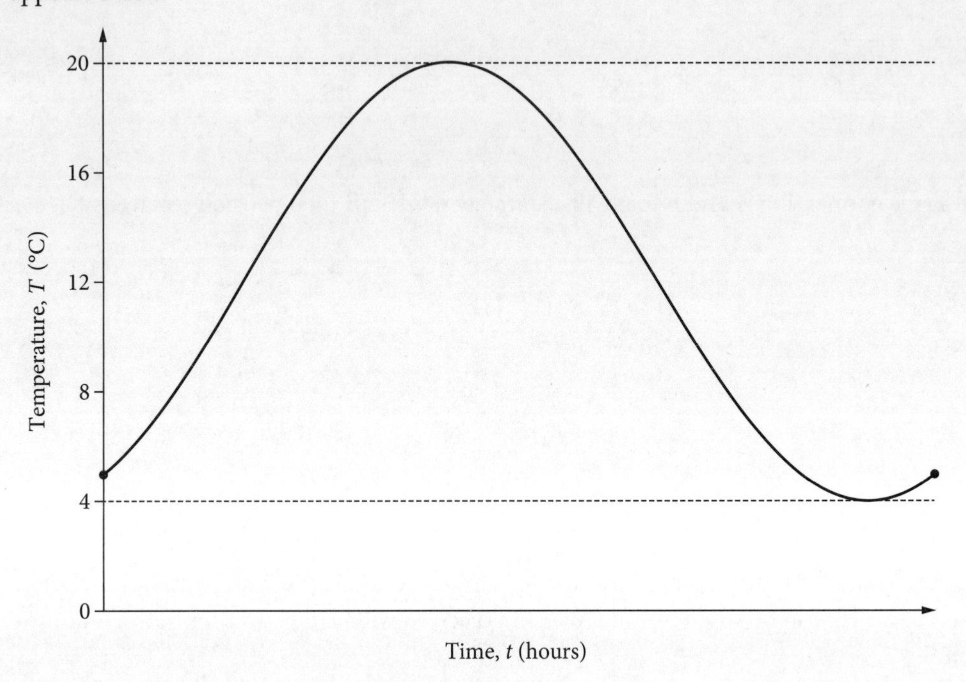

a Find the temperature at 9 am, correct to one decimal place. 1 mark

Question 25 continues on page 161

Question 25 (continued)

b At what time does the temperature first reach 16°C? 2 marks

c For what values of t is T decreasing? 2 marks

End of Question 25

Question 26 (5 marks)

The spinner below is spun twice and the 2 numbers spun are added together to make a sum, x.

The probability distribution table for this event appears below.

x	2	3	4	5	6
$p(x)$	$\frac{1}{4}$	$3k$	$\frac{5}{18}$	k	$\frac{1}{36}$

Find the standard deviation for this distribution, correct to two decimal places. 5 marks

Question 27 (6 marks)

The diagram shows the beginning of a series of concentric circles. The radius of the innermost circle is 1 unit. The radius of each circle, moving outwards, is 1 unit more than the previous circle.

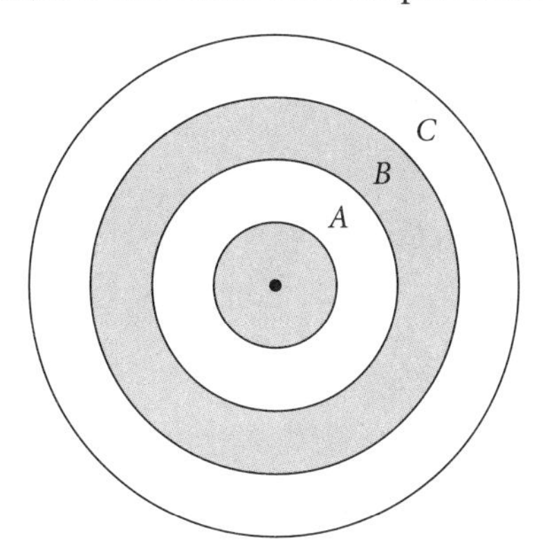

The ring-shaped area between two concentric circles is called an annulus.

The area of the annulus marked A is 3π square units. The area of the annulus marked B is 5π square units.

The area of the annulus marked C is 7π square units and so on.

a Find the exact area of the 10th annulus. 1 mark

b Show that the sum of the areas of the first n annuli is found by 2 mark

$$S_n = \pi n^2 + 2n\pi.$$

Question 27 continues on page 164

Question 27 (continued)

c What is the minimum number of annuli needed for a combined area of 300 units^2? 3 marks

Question 28 (6 marks)

An iron switches itself off after a period of inactivity. The temperature of the iron can be modelled by the equation

$$T = 22 + 158(1.2)^{-0.5t},$$

where T is the temperature of the iron, in °C, t minutes after it switches off.

a What is the temperature of the iron 6 minutes after it has switched off? Give your answer to the nearest degree. 1 mark

Question 28 continues on page 165

Question 28 (continued)

b At what rate is the temperature of the iron's surface changing after 6 minutes? Give your answer to one decimal place. 2 marks

c How long, to the nearest minute, will it take for the iron to cool to 36°C? 2 marks

d What value will the temperature of the iron approach? 1 mark

End of Question 28

9780170459235

Question 29 (6 marks)

Leonard and Penny borrow \$400 000 to buy an apartment in the city. Interest is charged each month on the outstanding balance at a rate of 6% per annum (0.5% per month).

Immediately after the interest is added, Leonard and Penny make a repayment of \$*M*. In an effort to repay the loan in 20 years, each subsequent repayment is 1% greater than the previous one.

Let A_n be the balance owing after the *n*th repayment has been made.

a Show that $A_2 = 400\,000(1.005^2) - M(1.005 + 1.01)$. 1 mark

b Show that $A_3 = 400\,000(1.005^3) - M[1.005^2 + (1.005)(1.01) + 1.01^2]$. 1 mark

Question 29 continues on page 167

Question 29 (continued)

c Find the value of M that will allow Leonard and Penny to successfully repay the loan in 20 years. 4 marks

End of Question 29

Question 30 (9 marks)

An open cone of radius r cm and height h cm is made from a sector of a circle.

The area of the sector is 300 cm^2.

The volume of a cone is given by the formula $V = \frac{1}{3}\pi r^2 h$.

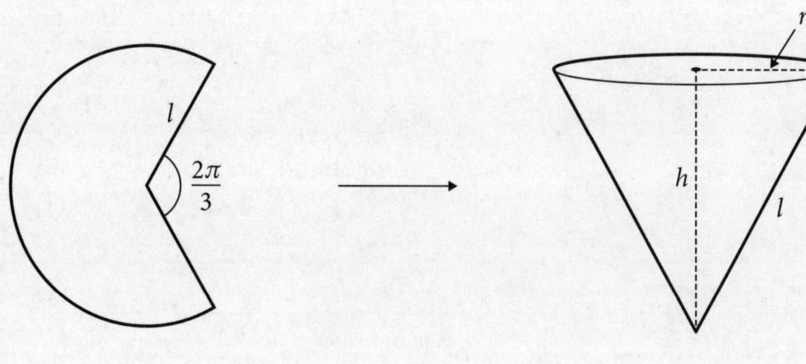

a Show that $l^2 = \frac{450}{\pi}$. 2 marks

b Show that the volume of the cone can be written as 2 marks

$$V = \frac{1}{3}r^2\sqrt{450\pi - \pi^2 r^2}.$$

Question 30 continues on page 169

Question 30 (continued)

c Show that $\frac{dV}{dr} = \frac{300\pi r - \pi^2 r^3}{\sqrt{450\pi - \pi^2 r^2}}$. 3 marks

d Find the value of r that maximises the volume of the cone. 2 marks

End of Question 30

Question 31 (5 marks)

The graph represents the function $y = a\sin bx$.

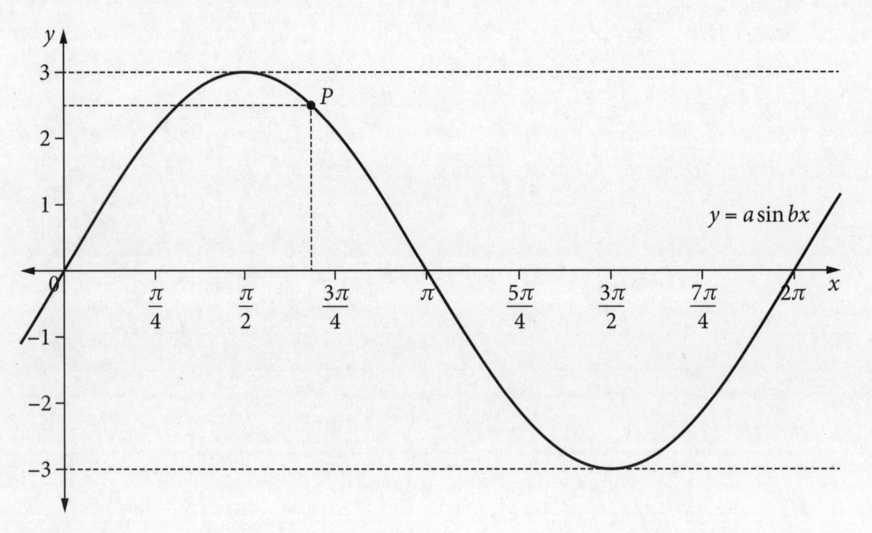

a Find the values of a and b. 2 marks

b Point $P\left(k, \frac{5}{2}\right)$ lies on the graph, as shown. 3 marks

Find the exact value of $\tan k$.

END OF PAPER

SECTION II EXTRA WRITING SPACE

WORKED SOLUTIONS

Section I (1 mark each)

Question 1

A $m = \frac{y_2 - y_1}{x_2 - x_1}$

$= \frac{3 - (-2)}{1 - 4}$

$= -\frac{5}{3}$

Year 9 question.

Question 2

D $\frac{a^4\sqrt{a}}{a^{-1}} = \frac{a^4 \times a^{\frac{1}{2}}}{a^{-1}}$

$= \frac{a^{\frac{9}{2}}}{a^{-1}}$

$= a^{\frac{11}{2}}$

Familiarity with all index laws is vital.

Question 3

B A discrete variable is one that can only take on specific numerical values; normally integers. Generally, it would be a quantity that was counted rather than measured. So, in this instance, it would be the number of visits to the cinema.

Know your definitions.
Mathematics is not just about calculations.

Question 4

D From the HSC exam reference sheet:

$y = \cos f(x)$ and $\frac{dy}{dx} = -f'(x)\sin f(x)$

It follows that if $y = \cos(x^2)$,
then $\frac{dy}{dx} = -2x\sin(x^2)$.

Typical HSC exam question. The mastery of the differentiation rules is essential to success in Maths Advanced.

Question 5

C $r = \frac{6\%}{12} = 0.5\%$, $n = 36$

Future value $= 200 \times 39.3361$
$= \$7867.22$

So interest $= \$7867.22 - (36 \times \$200)$
$= \$667.22$

Annuities is new to the course. Common with Maths Standard 2.

Question 6

Questions applying calculus to motion often provide information on a graph. Always check the graph first. Is it displacement (x) or velocity (v) vs time (t)? This will affect the answer completely.

B To answer this question, determine the sign of the velocity and acceleration at point P. Remembering that $v = \dot{x}$ (the derivative of displacement), since the graph is decreasing at P, velocity is negative and so the particle is moving left. As $a = \ddot{x}$, examine the concavity of the graph at P. Noticing it is concave down, acceleration is negative. The negative velocity is decreasing, so the *speed* is increasing.

So the correct option is B.

Band 5/6 question. There are quite a few connections that need to be made here that require more than a superficial grasp of this topic.

Question 7

C $2\log a = \log b - \log 3$

$\log a^2 = \log\left(\frac{b}{3}\right)$

$a^2 = \frac{b}{3}$

$a = \sqrt{\frac{b}{3}}$

Very common exam question. Know the logarithm laws.

Question 8

A $P(\text{John is absent}) = \frac{1}{20}$

$P(\text{Lucas is absent}) = \frac{1}{15}$

So $P(\text{both are absent}) = \frac{1}{20} \times \frac{1}{15}$

$= \frac{1}{300}$

Typical probability question. Common content with Maths Standard 2 🔗.

Question 9

A As it is a geometric sequence, the ratio of consecutive terms is constant.

So $\frac{10}{x} = \frac{x}{2}$

$x^2 = 20$

$x = \pm\sqrt{20}$

So x is possibly $-\sqrt{20}$.

Be able to answer a wide variety of questions on arithmetic and geometric series, not just substituting into the various formulas.

Question 10

B In order to be certain that $\tan\theta + \cos\theta < 0$, both $\tan\theta$ and $\cos\theta$ must be less than 0. This only happens in the 2nd quadrant (think ASTC) and so the statement is true for all values of θ in the domain $\left(\frac{\pi}{2}, \pi\right)$.

An unusual question examining the sign of the trigonometric ratios in each quadrant. Be flexible in your understanding and problem-solving.

Section II ($\checkmark$ = 1 mark)

Question 11 (2 marks)

$4x - xy^2 = x(4 - y^2)$ ✓

$= x(2 - y)(2 + y)$ ✓

Straightforward Year 9 algebra question. 2 marks and the word 'fully' means that 2 steps/lines are required in the working.

Question 12 (3 marks)

$x^2 - 4x + 1 = 0$

$x = \frac{-b \pm \sqrt{b^2 - 4ac}}{2a}$

$x = \frac{4 \pm \sqrt{4^2 - 4 \times 1 \times 1}}{2}$ ✓

$= \frac{4 \pm \sqrt{12}}{2}$

$= \frac{4 \pm 2\sqrt{3}}{2}$

$= 2 \pm \sqrt{3}$

So $a = 2$ ✓ and $b = 3$. ✓

Straightforward question although the way it is posed requires the simplification of the answer to express it in the form $a \pm \sqrt{b}$.

Question 13 (2 marks)

$y = 2x^2 - 3$

So $y' = 4x$.

When $x = 1$, $y' = 4$. ✓

$y - y_1 = m(x - x_1)$

$y + 1 = 4(x - 1)$ ✓

$y + 1 = 4x - 4$

$y = 4x - 5$

Typical tangent HSC question.

Question 14 (2 marks)

a $P(\text{no type B}) = 0.9 \times 0.9 \times 0.9$

$= 0.729$ ✓

b $P(\text{at least one type B}) = 1 - P(\text{no type B})$

$= 1 - 0.729$

$= 0.271$ ✓

Popular probability problem. Common with Maths Standard 2 🔗. The phrase 'at least one' is extremely common in probability questions.

Question 15 (2 marks)

$\sec\left(\frac{3\pi}{4}\right) = \frac{1}{\cos\left(\frac{3\pi}{4}\right)}$ ✓

Now $\cos\left(\frac{3\pi}{4}\right) = -\cos\left(\frac{\pi}{4}\right) = -\frac{1}{\sqrt{2}}$

So $\sec\left(\frac{3\pi}{4}\right) = \frac{1}{-\frac{1}{\sqrt{2}}} = -\sqrt{2}$ ✓

Year 11 work. Know the reciprocal ratios and the exact values for $\frac{\pi}{6}$, $\frac{\pi}{4}$ and $\frac{\pi}{3}$.

Question 16 (4 marks)

a $A = 9000$, $r = 0.95$

$T_n = ar^{n-1}$

So $T_3 = 9000(0.95^2)$ ✓

$= \$8122.50$

b $r = 0.95$, which is between −1 and 1, so the geometric series has a limiting sum.

$$S = \frac{a}{1-r} = \frac{9000}{1-0.95} = \$180\,000\text{, as required}$$ ✓

c $S_n = \frac{a(1-r^n)}{1-r}$

$S_4 = \frac{9000(1-0.95^4)}{1-0.95}$ ✓

$\approx \$33\,388.88$

Percentage of sales achieved in the first 4 months

$= \frac{\$33\,388.88}{\$180\,000} \times 100\%$

$\approx 18.5\%$ ✓

Applications of series are common in the HSC exam, using formulas that are found on the HSC exam reference sheet.

Question 17 (6 marks)

a The line l is $3x + 2y + 7 = 0$.

In gradient-intercept form, l is $y = -\frac{3}{2}x - \frac{7}{2}$.

So $m_l = -\frac{3}{2}$ and $m_{NP} = \frac{2}{3}$. ✓

$$y - y_1 = m(x - x_1)$$

$$y + 4 = \frac{2}{3}(x + 4)$$ ✓

$$3y + 12 = 2(x + 4)$$

$$3y + 12 = 2x + 8$$

$$0 = 2x - 3y - 4\text{, as required.}$$

It is important to not only know what information is needed but how to find it when it has not been specifically stated, in this case, the gradient of NP.

b P is the point of intersection of line l and NP. So solve simultaneously.

$3x + 2y + 7 = 0$ [1]

$2x - 3y - 4 = 0$ [2]

[1] × 2

[2] × 3

$6x + 4y + 14 = 0$ [3]

$6x - 9y - 12 = 0$ [4]

[3] − [4]

$13y + 26 = 0$

$y = -2$ ✓

Substitute in [1]:

$3x + 2(-2) + 7 = 0$

$3x - 4 + 7 = 0$

$3x = -3$

$x = -1$ ✓

So P is the point $(-1, -2)$.

Set your work out well. It should always be abundantly clear what you are doing at each step.

c By the distance formula:

$$d = \sqrt{(x_2 - x_1)^2 + (y_2 - y_1)^2}$$

$$= \sqrt{(-1-(-4))^2 + (-2-(-4))^2}$$

$$= \sqrt{3^2 + 2^2}$$

$$= \sqrt{13}$$ ✓

d $(x + 4)^2 + (y + 4)^2 = 13$ ✓

Note that the parts of this question are related, and the answer to one part is required for the next part. Always look for this.

Question 18 (5 marks)

Find the length of each unknown side.

By the cosine rule in ΔBCE,

$$BC^2 = 60^2 + 90^2 - 2 \times 60 \times 90 \times \cos 35°$$
$$= 2853.157\ldots$$
$$BC \approx 53.4\,\text{km.} \checkmark$$

By Pythagoras' theorem in ΔCDE,

$$CD^2 = 100^2 - 90^2$$
$$= 1900$$
$$CD \approx 43.6\,\text{km.} \checkmark$$

By the sine rule in ΔABE,

$$\frac{AE}{\sin 45°} = \frac{60}{\sin 80°} \checkmark$$
$$AE = \frac{60 \sin 45°}{\sin 80°}$$
$$\approx 43.1\,\text{km.} \checkmark$$

$$\text{Length of course} = 50 + 53.4 + 43.6 + 100 + 43.1$$
$$\approx 290\,\text{km} \checkmark$$

Straightforward question despite the lack of scaffolding (steps/hints). For a 5-mark trigonometry problem, spend some time planning your approach first. All concepts are common with Maths Standard 2.

Question 19 (3 marks)

a $300 000, 20 years

From the table, the monthly repayment is $1946.87. ✓

b Cost of the loan = $1946.87 × 20 × 12
= $467 248.80

So interest paid = $467 248.80 – $300 000
= $167 248.80 ✓

c For a 15-year term, the monthly repayment will be $2341.24.

Cost of the loan = $2341.24 × 15 × 12
= $421 423.20

So savings = $467 248.80 – $421 423.20
= $45 825.60 ✓

Question **19** is a typical HSC question with common content with Maths Standard 2.

Question 20 (2 marks)

$$\text{Area} \approx \frac{2}{2}\left[48 + 2(70 + 56 + 30) + 16\right] \checkmark$$
$$= 376 \text{ people} \checkmark$$

Unusual in that the information is provided graphically rather than as a function/equation.

Question 21 (5 marks)

a

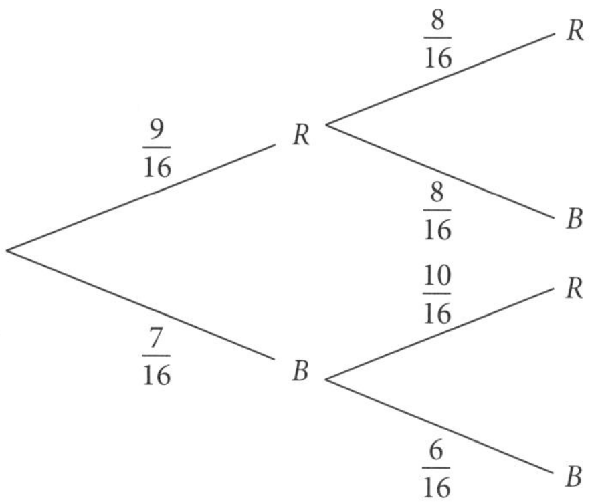

First marble: R $\left(\frac{9}{16}\right)$, B $\left(\frac{7}{16}\right)$

After R: R $\left(\frac{8}{16}\right)$, B $\left(\frac{8}{16}\right)$

After B: R $\left(\frac{10}{16}\right)$, B $\left(\frac{6}{16}\right)$

$$P(\text{first marble is blue}) = \frac{7}{16} \checkmark$$

This entire question is typical of a HSC exam question but with a twist. Except for part **b**, it is common with Maths Standard 2. Tree diagrams often feature in HSC exams. If the question makes a suggestion, as in this case, it is a good idea to use it, unless you are very confident it is not needed.

b $P(\text{second is blue} \mid \text{first is blue}) = \frac{6}{16} = \frac{3}{8} \checkmark$

A fairly simple conditional probability question.

c $P(\text{both marbles are red}) = \frac{9}{16} \times \frac{8}{16} = \frac{9}{32} \checkmark$

Straightforward use of the product rule of probability.

d $P(\text{marbles are different colours})$

$$= P(\text{RB}) + P(\text{BR})$$
$$= \left(\frac{9}{16} \times \frac{8}{16}\right) \times \left(\frac{7}{16} \times \frac{10}{16}\right) \checkmark$$
$$= \frac{71}{128} \checkmark$$

Question 22 (2 marks)

$f(x) = x^2 + 1$ and $g(x) = \sqrt{x-3}$.

$$\text{So } f(g(x)) = (\sqrt{x-3})^2 + 1$$
$$= x - 3 + 1$$
$$= x - 2 \checkmark$$

But since $g(x)$ is only defined over the domain $[3, \infty)$, this also becomes the domain of $f(g(x))$. So, the graph is:

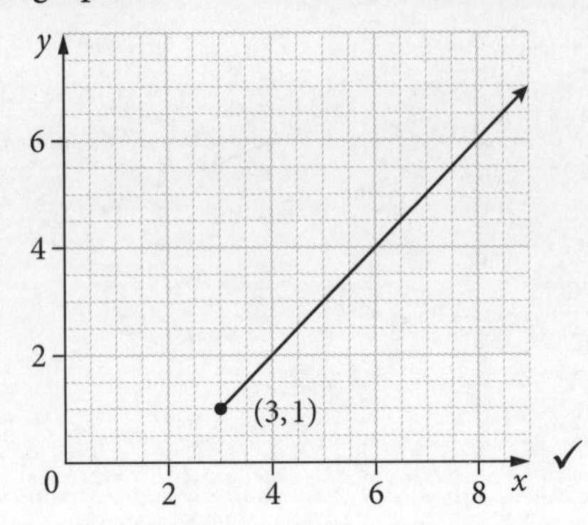

Composite functions is new content. A similar question appears in Question 17 of the NESA 2020 sample HSC exam (reproduced in Topic exam 1 on page 9 of this book).

Question 23 (4 marks)

a Required to prove $\cos^2\theta(\sec^2\theta - 1) = \sin^2\theta$.

$$\text{LHS} = \cos^2\theta(\sec^2\theta - 1)$$
$$= \cos^2\theta(\tan^2\theta) \checkmark$$
$$= \cos^2\theta\left(\frac{\sin^2\theta}{\cos^2\theta}\right)$$
$$= \sin^2\theta \checkmark$$
$$= \text{RHS}$$

Practise using the trigonometric identities.

b $4\cos^2\theta(\sec^2\theta - 1) = 1$

From part **a**,

$$4\sin^2\theta = 1 \checkmark$$
$$\sin^2\theta = \frac{1}{4}$$
$$\sin\theta = \pm\frac{1}{2}$$
$$\text{So } \theta = \frac{\pi}{6}, \frac{5\pi}{6}, \frac{7\pi}{6}, \frac{11\pi}{6} \checkmark$$

With trigonometric equations, make sure you include *all* solutions.

Question 24 (6 marks)

a

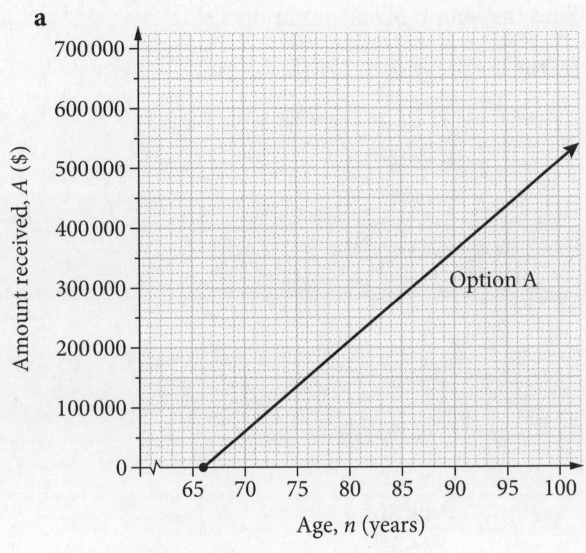

✓ graph starts at $n = 66$

✓ gradient is 15 000, goes through (76, 150 000)

Common content with Maths Standard 2.

b **i** Pension received $= 25 \times 10 \times \$760$
$= \$190\,000$ ✓

ii Each year, Surinder receives 25 payments of \$760, which totals \$19 000.

The number of years he has received this is his age beyond 70; that is $(n - 70)$.

So $A = 19\,000(n - 70)$. ✓

Interpretation of practical situations is important. Creating a formula from a worded problem is often tested in the HSC exam.

c

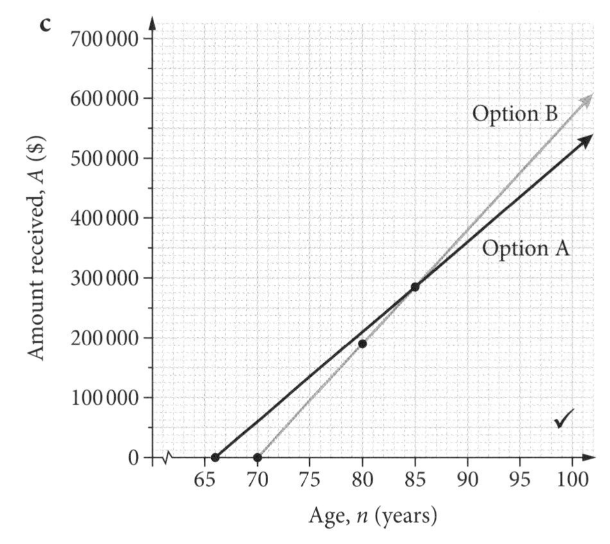

The graph for Option B starts at $n = 70$ and has a gradient of 19 000, so it goes through (80, 190 000). The point of intersection is at $n = 85$.

So, at age 85, Surinder's total pension received will be the same for both options. ✓

OR, using algebra:

$$\begin{aligned} 15000(n-66) &= 19000(n-70) \quad ✓ \\ 15(n-66) &= 19(n-70) \\ 15n - 990 &= 19n - 1330 \\ 4n &= 340 \\ n &= 85 \end{aligned}$$

So at age 85, Surinder's total pension received will be the same for both options. ✓

Graphical approach common with Maths Standard 2 and often tested in HSC exams.

Question 25 (5 marks)

a $T = 8\sin\left[\frac{\pi}{12}(t-4)\right] + 12$

Since time is measured from 6 am, 9 am is when $t = 3$.

$$\begin{aligned} \text{So } T &= 8\sin\left[\frac{\pi}{12}(3-4)\right] + 12 \\ &= 8\sin\left(-\frac{\pi}{12}\right) + 12 \\ &\approx 9.9°\text{C} \quad ✓ \end{aligned}$$

Make sure your calculator is in RAD mode for trigonometric functions.

b $T = 8\sin\left[\frac{\pi}{12}(t-4)\right] + 12$

When $T = 16$,

$$\begin{aligned} 16 &= 8\sin\left[\frac{\pi}{12}(t-4)\right] + 12 \quad ✓ \\ 4 &= 8\sin\left[\frac{\pi}{12}(t-4)\right] \\ \frac{1}{2} &= \sin\left[\frac{\pi}{12}(t-4)\right] \\ \frac{\pi}{12}(t-4) &= \frac{\pi}{6} \\ \frac{1}{12}(t-4) &= \frac{1}{6} \\ t-4 &= 2 \\ t &= 6 \end{aligned}$$

So the temperature first reaches 16°C at 12 noon. ✓

Application of trigonometric functions but a higher level of difficulty due to the function.

c Considering the equation

$T = 8\sin\left[\frac{\pi}{12}(t-4)\right] + 12$, T is a maximum when $\sin\left[\frac{\pi}{12}(t-4)\right] = 1$ as this is the maximum value of the sine function. This will occur when $\frac{\pi}{12}(t-4) = \frac{\pi}{2}$ as $\sin\left(\frac{\pi}{2}\right) = 1$.

Solve for t:

$$\begin{aligned} \frac{\pi}{12}(t-4) &= \frac{\pi}{2} \\ \frac{1}{12}(t-4) &= \frac{1}{2} \\ t-4 &= 6 \\ t &= 10 \quad ✓ \end{aligned}$$

The maximum occurs at 4 pm.

Since the period of the function is 24 hours, the minimum occurs 12 hours later; that is, at 4 am.

So, the function is decreasing for $10 < t < 22$ or between 4 pm and 4 am. ✓

There are a variety of approaches possible here. The deeper your understanding, the more working you save. Using properties of trigonometric functions is much quicker and easier than using calculus. Learn concepts, not only rules and procedures.

WORKED SOLUTIONS

Question 26 (5 marks)

$$\frac{1}{4} + 3k + \frac{5}{18} + k + \frac{1}{36} = 1$$
$$4k + \frac{5}{9} = 1$$
$$4k = \frac{4}{9}$$
$$k = \frac{1}{9} \checkmark$$

$$\mu = \left(2 \times \frac{1}{4}\right) + \left(3 \times \frac{1}{3}\right) + \left(4 \times \frac{5}{18}\right) + \left(5 \times \frac{1}{9}\right) + \left(6 \times \frac{1}{36}\right)$$
$$= \frac{1}{2} + 1 + \frac{10}{9} + \frac{5}{9} + \frac{1}{6}$$
$$= \frac{10}{3} \checkmark$$

$$\text{Var}(X) = E(X^2) - \mu^2$$
$$= \left(4 \times \frac{1}{4}\right) + \left(9 \times \frac{1}{3}\right) + \left(16 \times \frac{5}{18}\right) + \left(25 \times \frac{1}{9}\right) + \left(36 \times \frac{1}{36}\right) - \left(\frac{10}{3}\right)^2 \checkmark$$
$$= 1 + 3 + \frac{40}{9} + \frac{25}{9} + 1 - \frac{100}{9}$$
$$= \frac{10}{9} \checkmark$$

So $\sigma = \frac{\sqrt{10}}{3} \approx 1.05.$ ✓

Discrete probability distributions are new to the Advanced course. Expect it to be examined fairly regularly. This 5-mark question gives no clues and requires your deep knowledge of the topic. It applies the processes central to the topic, which should be practised extensively.

Question 27 (6 marks)

a The areas of the annuli moving outwards form the arithmetic sequence $3\pi, 5\pi, 7\pi, \ldots$

$$T_n = a + (n-1)d$$
$$T_{10} = 3\pi + (10-1)2\pi$$
$$= 21\pi \text{ square units} \checkmark$$

Application of a formula on the HSC exam reference sheet, but the use of π is uncommon.

b

$$S_n = \frac{n}{2}[2a + (n-1)d]$$
$$= \frac{n}{2}[2(3\pi) + (n-1)2\pi] \checkmark$$
$$= n[3\pi + n\pi - \pi]$$
$$= n[n\pi + 2\pi]$$
$$= \pi n^2 + 2n\pi, \text{ as required} \checkmark$$

The numbers may be unusual, but this should not be problematic as the process is straightforward. In a 'show that' question, you are given a 'target' to aim for.

9780170459235

c $\pi n^2 + 2n\pi \geq 300$

$\pi n^2 + 2n\pi - 300 \geq 0$ ✓

Solving $\pi n^2 + 2n\pi - 300 = 0$:

$$n = \frac{-b \pm \sqrt{b^2 - 4ac}}{2a}$$

$$= \frac{-2\pi \pm \sqrt{4\pi^2 + 1200\pi}}{2\pi}$$

But $n > 0$,

$$\therefore n = \frac{-2\pi + \sqrt{4\pi^2 + 1200\pi}}{2\pi} \text{ only. ✓}$$

So $n \approx 8.8$.

Considering the shape of the quadratic function, $f(x) = \pi n^2 + 2n\pi - 300$, the minimum value of n would be 9, meaning 9 annuli are required to give a total area of 300 units^2. ✓

Quite a complex question with a reliance on good algebraic skills as well as a knowledge of quadratic inequalities. Is there an easier way? What if we merged all of the annuli?

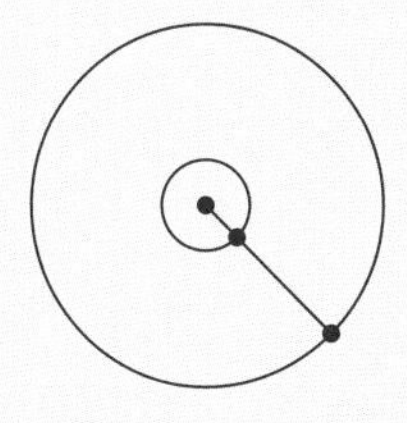

Remember, the inner circle has radius 1. Let the outer radius be r.

Investigate it. Think. This is what builds your skill.

Mathematics is 99% inspiration and 1% perspiration, not the other way around.

Question 28 (6 marks)

a $T = 22 + 158(1.2)^{-0.5t}$

When $t = 6$,

$$T = 22 + 158(1.2)^{-0.5(6)} \text{ ✓}$$
$$\approx 113°\text{C}$$

b $T = 22 + 158(1.2)^{-0.5t}$

$$\text{So } \frac{dT}{dt} = 158(-0.5)(1.2)^{-0.5t}(\ln 1.2) \text{ ✓}$$
$$= -79(1.2)^{-0.5t}(\ln 1.2)$$

When $t = 6$,

$$\frac{dT}{dt} = -79(1.2)^{-0.5(6)}(\ln 1.2)$$
$$\approx -8.3°\text{C per minute. ✓}$$

Higher-level question. Well-developed algebraic skills are required. Applying calculus to exponential functions where the base is not e is an important skill. See also Question 21 of the 2020 HSC exam.

c $T = 22 + 158(1.2)^{-0.5t}$

When $T = 36$,

$$36 = 22 + 158(1.2)^{-0.5t} \text{ ✓}$$
$$14 = 158(1.2)^{-0.5t}$$
$$\frac{7}{79} = (1.2)^{-0.5t}$$
$$\text{So } \ln\left(\frac{7}{79}\right) = \ln(1.2)^{-0.5t}$$
$$\ln\left(\frac{7}{79}\right) = (-0.5t)\ln(1.2)$$
$$-0.5t = \frac{\ln\left(\frac{7}{79}\right)}{\ln(1.2)}$$
$$t = \frac{-2\ln\left(\frac{7}{79}\right)}{\ln(1.2)} \text{ ✓}$$
$$= 26.5853\ldots$$
$$\approx 27 \text{ minutes}$$

The use of logarithms to solve equations is a skill that Mathematics Advanced students are expected to master.

d As $t \to \infty$, $(1.2)^{-0.5t} \to 0$.

So $T \to 22 + 158(0) = 22°\text{C}$ ✓

An understanding of the limiting value of an exponential function is required here.

Question 29 (6 marks)

a $A_1 = 400\,000(1.005) - M$

$$\begin{aligned} A_2 &= A_1(1.005) - 1.01M \\ &= [400\,000(1.005) - M](1.005) - 1.01M \\ &= 400\,000(1.005)^2 - 1.005M - 1.01M \\ &= 400\,000(1.005)^2 - M(1.005 + 1.01), \text{ as required. } \checkmark \text{ for process} \end{aligned}$$

Not your standard time payments question. The increase each month in the repayment significantly raises the difficulty of this question.

b
$$\begin{aligned} A_3 &= A_2(1.005) - 1.01^2M \\ &= [400\,000(1.005)^2 - 1.005M - 1.01M]1.005 - 1.01^2M \\ &= 400\,000(1.005)^3 - 1.005^2M - (1.005)(1.01)M - 1.01^2M \\ &= 400\,000(1.005)^3 - M[1.005^2 + (1.005)(1.01) + 1.01^2] \; \checkmark \text{ for process} \end{aligned}$$

Keep your eye out for patterns.

c Generalising this pattern, we get,

$A_n = 400\,000(1.005)^n - M[1.005^{n-1} + (1.005)^{n-2}(1.01) + (1.005)^{n-3}(1.01)^2 + \cdots + (1.01)^{n-1}]$ ✓

The expression in the square brackets is a geometric series with $a = (1.005)^{n-1}$, and n terms.

$$A_n = 400\,000(1.005)^n - M\left[\frac{1.005^{n-1}\left[\left(\frac{1.01}{1.005}\right)^n - 1\right]}{\frac{1.01}{1.005} - 1}\right] \checkmark$$

$$= 400\,000(1.005)^n - M\left[\frac{1.005^{n-1}\left[\left(\frac{1.01}{1.005}\right)^n - 1\right]}{\frac{0.005}{1.005}}\right]$$

$$= 400\,000(1.005)^n - M\left[1.005^{n-1} \times \left[\left(\frac{1.01}{1.005}\right)^n - 1\right] \times \frac{1.005}{0.005}\right]$$

$$= 400\,000(1.005)^n - 200M\left[1.005^n\left(\frac{1.01^n}{1.005^n} - 1\right)\right]$$

So $A_n = 400\,000(1.005)^n - 200M[1.01^n - 1.005^n]$. ✓

If the loan is repaid in 20 years, $A_{240} = 0$.

$$\begin{aligned} 0 &= 400\,000(1.005)^{240} - 200M[1.01^{240} - 1.005^{240}] \\ &= 2000(1.005)^{240} - M[1.01^{240} - 1.005^{240}] \\ M[1.01^{240} - 1.005^{240}] &= 2000(1.005)^{240} \\ M &= \frac{2000(1.005)^{240}}{1.01^{240} - 1.005^{240}} \\ &= 873.134\,274\,9, \text{ by calculator} \\ &\approx \$873.13 \; \checkmark \end{aligned}$$

Quite a difficult question and one that is definitely aimed at Band 6 students. Sophisticated algebraic skills needed. Again, the key is the recognition of patterns. Notice in the initial equation for A_n, the sum of the indices in each term of the series is $n - 1$.

Don't skip steps in your working. This is how mistakes happen. Notice also the explanatory lines in the solution. Tell the marker what you're doing and why you're doing it.

WORKED SOLUTIONS

Question 30 (9 marks)

a The angle inside the sector is

$2\pi - \frac{2\pi}{3} = \frac{4\pi}{3}$. ✓

Now, $A = \frac{1}{2}r^2\theta$

$$\text{So } 300 = \frac{1}{2}l^2\left(\frac{4\pi}{3}\right)$$

$$1800 = 4\pi l^2$$

$$l^2 = \frac{1800}{4\pi}$$

$$= \frac{450}{\pi}, \text{ as required. } ✓$$

Maximisation and minimisation (optimisation) are concepts that give students the most difficulty in the Maths Advanced course due to the large number of scenarios possible. Regardless, well-developed algebra skills are always needed.

b $V = \frac{1}{3}\pi r^2 h$

By Pythagoras' theorem,

$$h^2 = l^2 - r^2$$

$$= \frac{450}{\pi} - r^2$$

$$h = \sqrt{\frac{450}{\pi} - r^2} ✓$$

$$\text{So } V = \frac{1}{3}\pi r^2\sqrt{\frac{450}{\pi} - r^2}$$

$$= \frac{1}{3}\sqrt{\pi^2 r^2}\sqrt{\frac{450}{\pi} - r^2}$$

$$= \frac{1}{3}r^2\sqrt{450\pi - \pi^2 r^2}, \text{ as required. } ✓$$

Invariably, part **a** will be needed in part **b**. Maximisation and minimisation questions invariably 'build' as you move through the parts. There's a reason each part of the question asks you to find something and gives you the answer.

c $V = \frac{1}{3}r^2\sqrt{450\pi - \pi^2 r^2}$

$$= \frac{1}{3}r^2(450\pi - \pi^2 r^2)^{\frac{1}{2}}$$

$$\frac{dV}{dr} = \frac{1}{3}r^2 \times \frac{1}{2}\left(450\pi - \pi^2 r^2\right)^{-\frac{1}{2}} \times (-2\pi^2 r) + \left(450\pi - \pi^2 r^2\right)^{\frac{1}{2}} \times \frac{2}{3}r ✓$$

$$= \frac{2}{3}r\sqrt{450\pi - \pi^2 r^2} - \frac{2\pi^2 r^3}{6\sqrt{450\pi - \pi^2 r^2}} ✓$$

$$= \frac{2r(450\pi - \pi^2 r^2)}{3\sqrt{450\pi - \pi^2 r^2}} - \frac{\pi^2 r^3}{3\sqrt{450\pi - \pi^2 r^2}}$$

$$= \frac{900\pi r - 2\pi^2 r^3 - \pi^2 r^3}{3\sqrt{450\pi - \pi^2 r^2}}$$

$$= \frac{900\pi r - 3\pi^2 r^3}{3\sqrt{450\pi - \pi^2 r^2}}$$

$$= \frac{300\pi r - \pi^2 r^3}{\sqrt{450\pi - \pi^2 r^2}}, \text{ as required } ✓$$

An excellent example of how precise and perfect your algebra must be to solve such problems.

This is a 'show that' question; you need to convince the marker that you would have arrived at the answer even if it were not provided. So, don't skip steps. 'Fudging' (bluffing) is always obvious.

d V is a maximum when $\frac{dV}{dr} = 0$.

$$0 = \frac{300\pi r - \pi^2 r^3}{\sqrt{450\pi - \pi^2 r^2}}$$
$$= 300\pi r - \pi^2 r^3$$
$$= 300 - \pi r^2$$
$$r^2 = \frac{300}{\pi}$$
$$r = \sqrt{\frac{300}{\pi}} \checkmark$$

r	9	$\sqrt{\frac{300}{\pi}}$	10
$\frac{dV}{dr}$	51.9	0	−21.5

So $r = \sqrt{\frac{300}{\pi}}$ maximises V. ✓

The equation appears difficult at first but fractions can only be zero when the numerator is zero.
Dividing both sides by r is fine since r is clearly not 0.
Again, a table of values is used to test whether V is a maximum at $r = \sqrt{\frac{300}{\pi}}$.
Clearly, using the second derivative here would have been impractical.

Question 31 (5 marks)

a This sine graph has a period of 2π and an amplitude of 3, so its equation is $y = 3\sin x$.

$a = 3$ ✓

$b = 1$ ✓

b $y = 3\sin x$

Substitute $\left(k, \frac{5}{2}\right)$.

$$\frac{5}{2} = 3\sin k \checkmark$$
$$\sin k = \frac{5}{6}$$

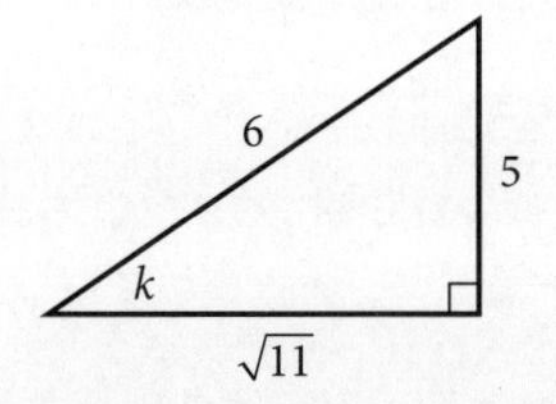

Let the base of the triangle be b.

$$b^2 = 6^2 - 5^2$$
$$b^2 = 11$$
$$b = \sqrt{11}$$

So $\tan k = \frac{5}{\sqrt{11}}$. ✓

But k is in 2nd quadrant so $\tan k < 0$.

So $\tan k = -\frac{5}{\sqrt{11}}$. ✓

Band 6-style question. Don't expect that every question will be posed in a familiar manner.
To achieve at the highest level, a certain degree of intuition and problem-solving is required, which is not always easy under exam conditions.

The 2020 Mathematics Advanced HSC Exam Worked Solutions

The 2020 HSC exam and other past HSC papers can be downloaded from the NESA website (www.educationstandards.nsw.edu.au) by selecting 'Year 11 – Year 12', 'HSC exam papers'. NESA marking feedback and guidelines can also be found there.

9780170459235

WORKED SOLUTIONS

Section I (1 mark each)

Question 1

D $2x - 3 \geq 0$ for domain

$$2x \geq 3$$

$$x \geq \frac{3}{2}.$$

Straightforward question.

Question 2

B When $y = x^3$ is changed to $y = (x-2)^3$, the graph is translated 2 units **right**.

Changing $y = (x-2)^3$ to $y = (x-2)^3 + 5$ translates the graph 5 units **up**.

An understanding of the connection between changes to the function and its graph is vital.

Question 3

A It is not possible to compare scores from different data sets until they are expressed as z-scores.

French: $\frac{82-70}{8} = 1.5$

Commerce: $\frac{80-65}{5} = 3$

Music: $\frac{74-50}{12} = 2$

The best result is Commerce and the worst is French.

Common content with Maths Standard 2.

Question 4

B $\int (e + e^{3x})\,dx = ex + \frac{1}{3}e^{3x} + c$

Integration of an exponential function. Remember: that first e is a constant.

Question 5

C $a = -1 < 0$

The graph must be concave down, eliminating options A and B.

The equation of the axis of symmetry is $y = -\frac{b}{2a}$.

Since $a = -1$ and $b > 0$, $-\frac{b}{2a} > 0$, so the axis of symmetry is to the right of the y-axis, eliminating D.

This content is first covered in Year 10. Again, connecting the features of a graph with its formula.

Question 6

B The centre is $y = 5$.

The amplitude is 2.

So the range is $[5-2, 5+2] = [3, 7]$.

Be familiar with trigonometric graphs.

Question 7

A $\int_8^{12} f(x)\,dx = 0$, as the two triangular areas cancel each other out.

So $\int_0^{12} f(x)\,dx = \int_0^8 f(x)\,dx$.

Calculating the area under the curve from 0 to 8, we get a long rectangle and a semicircle:

$$\text{Integral} = (8 \times 3) + \left(\frac{1}{2} \times \pi \times 2^2\right)$$

$$= 24 + 2\pi$$

Connecting the integral with the area under a graph. Also tests the idea that areas under the x-axis are 'negative' when the integral is evaluated.

Question 8

A

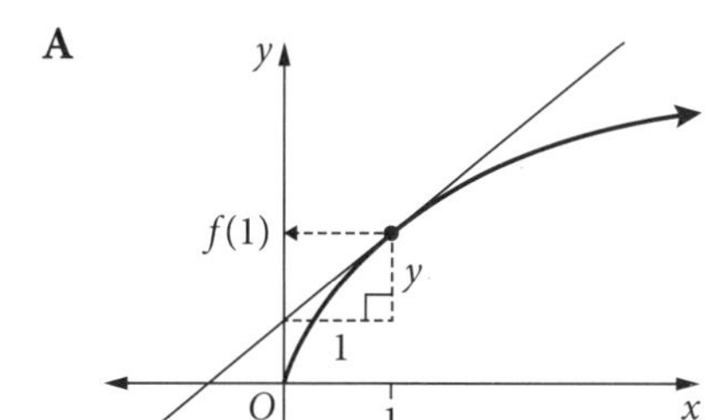

The value of the second derivative relates to the concavity of the graph. Since it is concave down, we know that $f''(1) < 0$. This rules out options C and D.

Rule a tangent to the curve at the point where $x = 1$. The gradient of this tangent is the value of $f'(1)$. This can be found using the small triangle drawn and applying $\frac{\text{rise}}{\text{run}} = \frac{y}{1} = y$.

$f(1) > y$, so $f(1) > f'(1)$.

This eliminates B, making A the correct choice.

Difficult question. It is, however, quite easy to determine that $f''(1) < 0$, which eliminates two options. Connects the idea that the value of the derivative at any point corresponds to the gradient of the tangent at that point. These connections are extremely important in Mathematics Advanced.

Question 9

C This question requires knowledge of the following overlay of diagrams.

The distribution is normal, so the mean, median (and mode) are all equal.

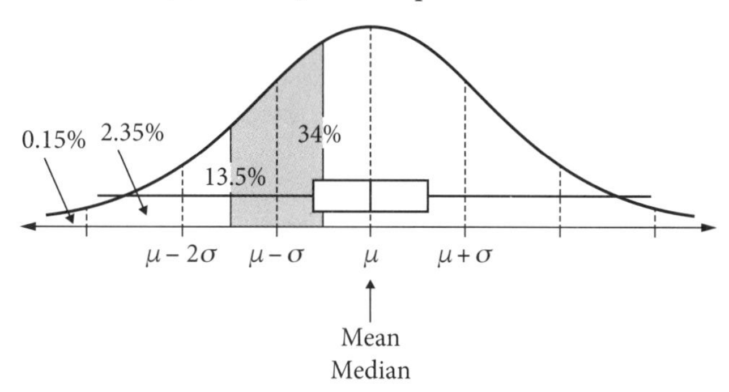

Using the combined knowledge of the percentages for both the normal distribution and box-and-whisker plots, the weight of the melon will lie somewhere on the left whisker (25%) of the box plot. It's also above the 10th percentile, so the region starts somewhere between $\mu - 2\sigma$ and $\mu - \sigma$.

C gives the correct region.

Common content with Maths Standard 2 but a difficult example of that content.

Question 10

D $h(x) = f(g(x))$

So $h'(x) = g'(x)f'(g(x))$, by the chain rule.

Stationary points when $h'(x) = 0$, so solve.

$$0 = g'(x)f'(g(x))$$

So either $g'(x) = 0$ OR $f'(g(x)) = 0$

So $x = 3$ due to the stationary point on $g(x)$.

Stationary point on $f(x)$ at $x \approx 0.8$.

We require $g(x) \approx 0.8$, which occurs twice: draw a horizontal line $y = 0.8$ and it will produce two points of intersection on $y = g(x)$.

So there is a total of 3 stationary points on $y = h(x)$.

Very difficult question and definitely targeting Band 6 students. Not only are composite functions new content but it is being applied in an unusual way here.

Section II (✓ = 1 mark)

Question 11 (4 marks)

a The graph for Tank A is a line that has a vertical intercept of 1000 and a gradient of –20. It can be drawn by graphing $V = 1000 - 20t$ or by noticing that the volume will be 0 after $\frac{1000}{20} = 50$ minutes, so the line will run from (0, 1000) to (50, 0).

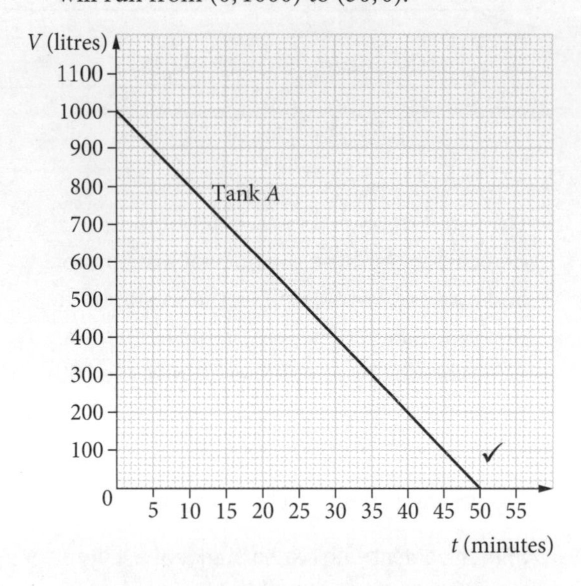

b The graph for Tank B is a line that starts from (15, 0) with a gradient of 30. It can be drawn using a table of values (every 10 minutes the volume will rise by 30 × 10 = 300 L)

t	15	25	35	45	55
V	0	300	600	900	1200

or by noticing that the volume will be 30 × 30 = 900 L after another 30 minutes (at $t = 45$), so the line will run from (15, 0) to (45, 900).

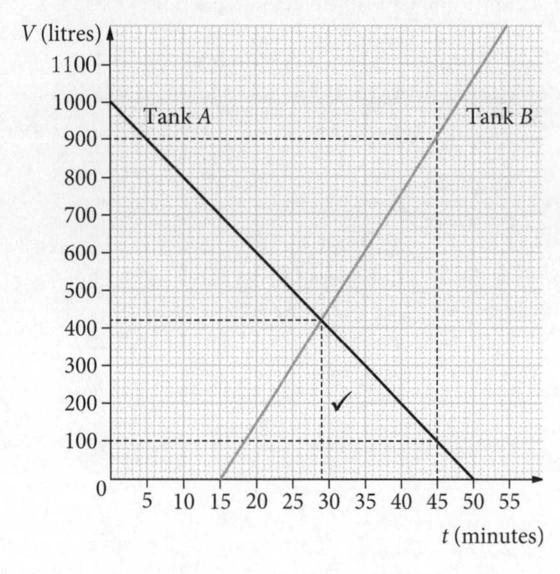

Point of intersection of the 2 graphs is (29, 420) (see black and grey lines in previous diagram).

Both tanks contain the same volume of water at 29 minutes. ✓

c By guess-and-check, when $t = 45$, volume of Tank A = 100 L and volume of Tank B = 900 L, giving a total of 1000 L (see dashed lines in the diagram in part **b**).

The total volume is 1000 L when $t = 45$ minutes. ✓

All of Question **11** was also in the Maths Standard 2 exam. Make sure you bring a ruler to the exam and use it to draw accurate graphs. Also, be prepared to graph lines for which the equation is not given in the question. Instead, the question describes a linear relationship in words.

Question 12 (3 marks)

$T_n = a + (n - 1)d$

$a = 4 \quad d = 6$

$1354 = 4 + (n - 1)6$ ✓

So $6(n - 1) = 1350$

$n - 1 = 225$

$n = 226$ ✓

$$S_n = \frac{n}{2}(a + l)$$

$$\text{So } S_{226} = \frac{226}{2}(4 + 1354)$$

$$= 153\,454$$ ✓

Straightforward question applying formulas from the HSC exam reference sheet. HSC markers reported common student errors in algebra and calculation, including answers for n that were not whole numbers.

Question 13 (2 marks)

$$\int_0^{\frac{\pi}{4}} \sec^2 x \, dx = [\tan x]_0^{\frac{\pi}{4}}$$ ✓

$$= \tan\left(\frac{\pi}{4}\right) - \tan 0$$

$$= 1 - 0$$

$$= 1$$ ✓

Straightforward question. Recognise the exact trigonometric ratios for values in radians.

Question 14 (5 marks)

a Total studying History or Geography
$= 40 - 7 = 33$

Students studying either $= 20 + 18 = 38$
Students studying both $= 38 - 33 = 5$

History only $= 20 - 5 = 15$
Geography only $= 18 - 5 = 13$

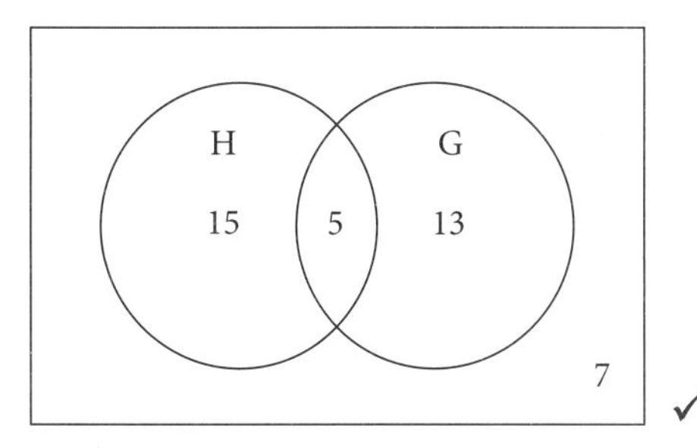

✓

$P(\text{both}) = \frac{5}{40} = \frac{1}{8}$ ✓

If the question suggests drawing a Venn diagram, it is always wise to take the advice. Many students did not check that their Venn diagram numbers added up to 40. Make sure you read all questions carefully and check that you answer the question asked.

b $P(\text{G but not H}) = \frac{13}{18}$ ✓

Make sure you understand the idea of conditional probability, and how the sample space (the denominator in the probability) is reduced from 40 to 18.
Students who used their Venn diagrams did well, while those who memorised a formula did not fare as well.

c $P(\text{H},\tilde{\text{H}}) = \frac{20}{40} \times \frac{20}{39}$ ✓

$= \frac{10}{39}$ ✓

Parts **b** and **c** are straightforward questions. Learn the meanings of 'without replacement', 'and' and 'or', and use them to solve two-stage probability problems.

Question 15 (5 marks)

a $\angle APB = 100° - 35°$
$= 65°$ ✓

b $AB^2 = 7^2 + 9^2 - 2 \times 7 \times 9 \cos 65°$ ✓
$= 130 - 126\cos 65°$

So $AB = 8.7607\ldots$
$\approx 8.76\,\text{km}$ ✓

c $\cos\angle PAB = \frac{7^2 + 8.7607\ldots^2 - 9^2}{2 \times 7 \times 8.7607\ldots}$
$= 0.3648\ldots$
$\angle PAB = 69°$ ✓

So bearing $= (180° + 35°) - 69°$
$= 146°$ ✓

Parts **a** to **c** are straightforward and common content with Maths Standard 2. Many students have trouble working with bearings. Practise calculations with bearings and the sine/cosine rules as they occur frequently in exams. Make sure your calculator is in DEG mode, use the reference sheet and show all of your working.

Question 16 (4 marks)

$y = -x^3 + 3x^2 - 1$

So $y' = -3x^2 + 6x$. ✓

Stationary points at $y' = 0$.

$0 = -3x^2 + 6x$
$= x^2 - 2x$
$= x(x - 2)$

So $x = 0$ or 2.

When $x = 0$, $y = -0^3 + 3(0^2) - 1$
$= -1 \quad (0, -1)$

When $x = 2$, $y = -2^3 + 3(2^2) - 1$
$= 3 \quad (2, 3)$

Determine the nature using y''.

$y'' = -6x + 6$
When $x = 0$, $y'' = 6$
> 0.

There is a minimum turning point at $(0, -1)$.

When $x = 2$, $y'' = -6$
< 0.

There is a maximum turning point at $(2, 3)$. ✓

Point of inflection at $y'' = 0$.

$0 = -6x + 6$
So $x = 1$.

When $x = 1$, $y = -1^3 + 3(1^2) - 1$
$= 1$

There is a possible point of inflection at $(1, 1)$.

x	$\frac{1}{2}$	1	$1\frac{1}{2}$
y''	3	0	-3

Concavity changes and so $(1, 1)$ is confirmed as a point of inflection. ✓

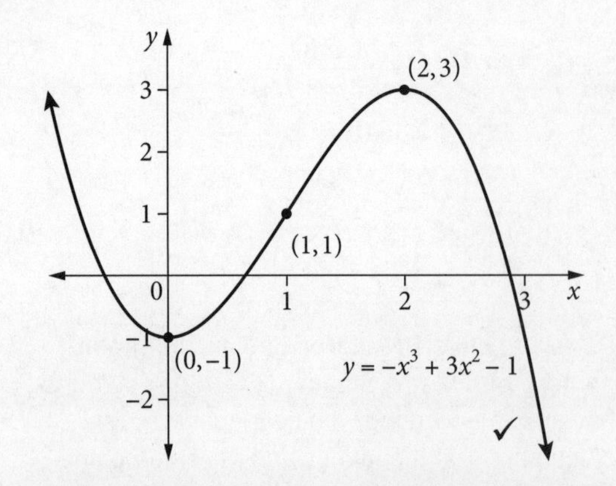

4-mark curve-sketching question that's almost always in the HSC exam. Don't forget to determine the nature of stationary points and test concavity on both sides of a point of inflection. Draw large diagrams, clearly labelled with even axes.

Question 17 (2 marks)

$$\int \frac{x}{4+x^2}\,dx = \frac{1}{2}\int \frac{2x}{4+x^2}\,dx$$
$$= \frac{1}{2}\ln(4+x^2)+c \quad ✓✓$$

Absolute value brackets are not needed because $4 + x^2$ is positive. Make sure you can distinguish between fractional functions that when integrated are logarithmic functions, such as $\left(\frac{2x}{4+x^2}\right)$, and those that are not, such as $\left(\frac{2}{4+x^2}\right)$. Make sure you insert brackets for logarithmic expressions: $\ln 4 + x^2$ is not the same as $\ln(4 + x^2)$.

Question 18 (3 marks)

a $$\frac{d}{dx}\left(e^{2x}(2x+1)\right) = e^{2x}(2) + (2x+1)2e^{2x} \quad ✓$$
$$= 2e^{2x}(1 + 2x + 1)$$
$$= 4e^{2x}(x+1) \quad ✓$$

Straightforward application of the product rule. Include brackets in your working to maintain clarity and minimise error.

b Since $\frac{d}{dx}\left(e^{2x}(2x+1)\right) = 4e^{2x}(x+1)$,

we can say that

$$\int 4e^{2x}(x+1)\,dx = e^{2x}(2x+1)+c$$

So $\int e^{2x}(x+1)\,dx = \frac{1}{4}e^{2x}(2x+1)+c.$ ✓

The inverse connection between differentiation and integration is often examined. Look at opportunities where the answer to part **b** relies on the answer to part **a**. In this question, 'hence' indicates this. Part **b** was worth one mark, but often students ignored their workings and answer to part **a** and did unnecessary working in part **b**.

Question 19 (2 marks)

$$\text{LHS} = \sec\theta - \cos\theta$$
$$= \frac{1}{\cos\theta} - \cos\theta$$
$$= \frac{1}{\cos\theta} - \frac{\cos^2\theta}{\cos\theta}$$
$$= \frac{1-\cos^2\theta}{\cos\theta} \quad ✓$$
$$= \frac{\sin^2\theta}{\cos\theta}$$
$$= \sin\theta \times \frac{\sin\theta}{\cos\theta}$$
$$= \sin\theta\tan\theta$$
$$= \text{RHS} \quad ✓$$

Straightforward application of trigonometric identities. Show every step and aim to start with the LHS and end with the RHS (or the other way around).

Question 20 (2 marks)

$$\text{Area} \approx \frac{\frac{1}{60}}{2}[60 + 2(55 + 65 + 68 + 70) + 67] \quad ✓$$
$$= 5.358\ldots$$
$$\approx 5.4\,\text{km} \quad ✓$$

Trapezoidal rule questions are almost always in the HSC exam. (See HSC exam topic grid on page 50). This one, however, gives you a table of values rather than a function, so the function values (v) are already given. Common student errors include not knowing how many function values to use and substituting t-values rather than v-values into the rule.

Question 21 (6 marks)

a $T = 25 + 70(1.5)^{-0.4t}$

When $t = 4$,

$$T = 25 + 70(1.5)^{-0.4(4)}$$
$$\approx 61.6°\text{C}. \quad ✓$$

Straightforward substitution that was well done.

b $T = 25 + 70(1.5)^{-0.4t}$

$$\text{So } \frac{dT}{dt} = 70(-0.4)(1.5)^{-0.4t}(\ln 1.5)$$
$$= -28(1.5)^{-0.4t}(\ln 1.5) \quad ✓$$

When $t = 4$,

$$\frac{dT}{dt} = -28(1.5)^{-0.4(4)}(\ln 1.5)$$
$$\approx -5.9°\text{C per minute}. \quad ✓$$

Many students could not differentiate exponential functions where the base was not e, apply the theory to a real-life problem, or realise that 'rate of change' (instantaneous, not average) meant the derivative. Remember to use or memorise the standard integrals on the reference sheet.

c $T = 25 + 70(1.5)^{-0.4t}$

When $T = 55$:

$$55 = 25 + 70(1.5)^{-0.4t}$$
$$30 = 70(1.5)^{-0.4t}$$
$$\frac{3}{7} = (1.5)^{-0.4t} \checkmark$$

So $\ln\left(\frac{3}{7}\right) = \ln(1.5)^{-0.4t}$.

$$\ln\left(\frac{3}{7}\right) = (-0.4t)\ln(1.5)$$
$$-0.4t = \frac{\ln\left(\frac{3}{7}\right)}{\ln(1.5)}$$
$$t = -\frac{\ln\left(\frac{3}{7}\right)}{0.4\ln(1.5)} \checkmark$$
$$= 5.2242\ldots \text{ minutes} \checkmark$$
$$\approx 5 \text{ minutes } 13 \text{ seconds}$$

Solving exponential equations is an essential skill for Mathematics Advanced students. When the base is not e, the change of base law can be applied. Check that your time answer makes sense: no negative or very big answers!

Question 22 (4 marks)

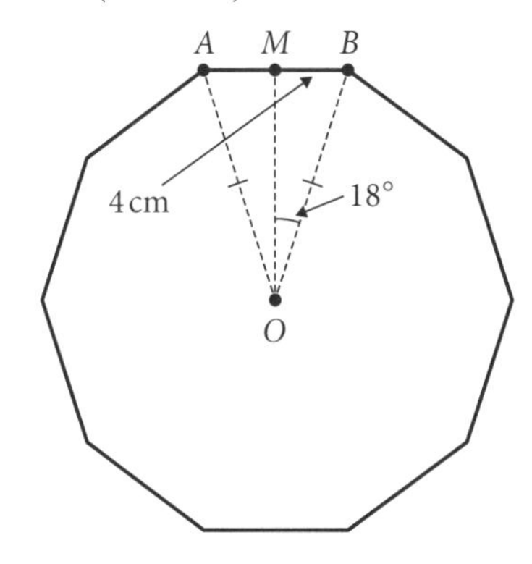

As the perimeter of the regular decagon is 80 cm, it can be determined that $AB = 8$ cm. ✓

Let M be the midpoint of AB. Then $MB = 4$ cm, as shown.

$\angle AOB = 360° \div 10 = 36°$. Therefore, $\angle MOB = 18°$.

Considering ΔMOB, $\tan 18° = \frac{4}{OM}$.

So $OM = \frac{4}{\tan 18°}$. ✓

$$\text{Area of } \Delta AOB = \frac{1}{2} \times 8 \times \frac{4}{\tan 18°}$$
$$= \frac{16}{\tan 18°} \checkmark$$

$$\text{So area of the decagon} = 10 \times \frac{16}{\tan 18°}$$
$$\approx 492.4 \text{ cm}^2 \checkmark$$

Although this is a complex 4-mark question with no parts/clues, all concepts used were covered in Years 9 and 10. The method shown is only one way to arrive at the answer, but possibly the simplest. There are many different methods possible, such as the sine rule; one of the benefits of a lack of scaffolding (steps and hints towards the solution) in a question.

Question 23 (4 marks)

a $\int_0^k \sin x \, dx = 1$ ✓

$$[-\cos x]_0^k = 1$$
$$-\cos k + \cos 0 = 1$$
$$-\cos k = 0$$
$$\cos k = 0$$
$$k = \frac{\pi}{2} \checkmark$$

Continuous probability distributions are new to the course. Practise answering as many different styles of questions as possible to be thoroughly prepared. This question involved integrating $\sin x$ and solving a trigonometric equation in *radians*, not degrees. Even though the integral can be found on the HSC exam reference sheet, many students wrote that the integral of $\sin x$ was $\cos x$ (the *derivative* of $\sin x$ is $\cos x$).

b $P(X \le 1) = \int_0^1 \sin x \, dx$ ✓

$$= [-\cos x]_0^1$$
$$= -\cos 1 + \cos 0$$
$$= 1 - \cos 1$$
$$\approx 0.4597 \checkmark$$

cos 1 means the cosine of 1 *radian* and not cos 1°. Degrees are *never* used with trigonometric functions, only in measurement problems for finding lengths, angles and areas. This question involved forming and evaluating the definite integral of $\sin x$ in radians.

WORKED SOLUTIONS

Question 24 (3 marks)

$$x^2 - 6x + y^2 + 4y - 3 = 0$$
$$x^2 - 6x + 9 + y^2 + 4y + 4 = 3 + 9 + 4$$
$$(x-3)^2 + (y+2)^2 = 16 \checkmark$$

This circle has its centre at $(3,-2)$ and a radius of 4.

Upon reflection in the x-axis, the centre moves to $(3,2)$ and the radius remains unchanged. ✓

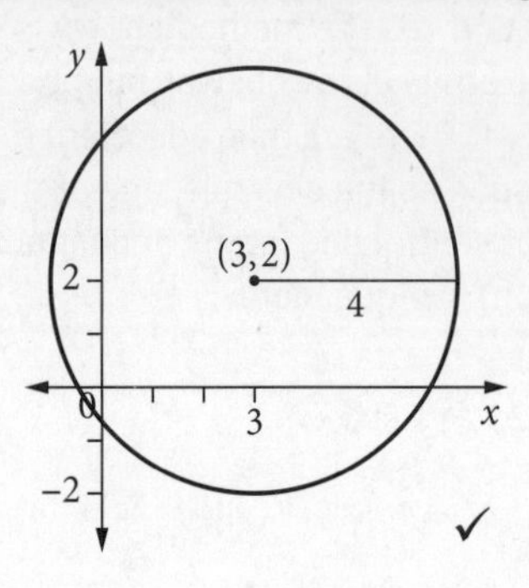

✓

It is quite common to be asked to find the centre and radius of a circle by completing the square. Make sure you can connect transformations and the changes that happen in the formula. The links between algebra and geometry are key in the Mathematics Advanced course. Again, bring a ruler to aid accurate graphing with well-labelled axes and features. Common errors were stating that the radius was 16, not 4, and forgetting to sketch the circle.

Question 25 (7 marks)

a Area $= xy + \frac{1}{4}\pi x^2$

$$36 = xy + \frac{\pi x^2}{4}$$
$$144 = 4xy + \pi x^2$$
$$4xy = 144 - \pi x^2$$

So $y = \frac{144 - \pi x^2}{4x}$. ✓

$P = 2x + 2y + \frac{1}{4}(2\pi x)$ ✓

$$= 2x + \frac{\pi x}{2} + 2y$$
$$= 2x + \frac{\pi x}{2} + 2\left(\frac{144 - \pi x^2}{4x}\right)$$
$$= 2x + \frac{\pi x}{2} + 2\left(\frac{36}{x} - \frac{\pi x}{4}\right)$$
$$= 2x + \frac{\pi x}{2} + \frac{72}{x} - \frac{\pi x}{2}$$
$$= 2x + \frac{72}{x} \checkmark$$

Optimisation questions have appeared in all HSC papers for many years. (See HSC exam topic grid on page 37). Examining the expression P, we see that it is in terms of x. Therefore, the fact given (that area is 36 m^2) must be used to establish an expression for y in terms of x, in order to eliminate y from the expression for P; a method very similar to the substitution method for solving simultaneous equations.

b Perimeter is minimised when $\frac{dP}{dx} = 0$.

$P = 2x + 72x^{-1}$

So $\frac{dP}{dx} = 2 - 72x^{-2}$. ✓

$$0 = 2 - \frac{72}{x^2}$$
$$\frac{72}{x^2} = 2$$
$$x^2 = 36$$
$$x = 6 \quad (x > 0) \checkmark$$

$$\frac{d^2P}{dx^2} = 144x^{-3}$$

When $x = 6$,

$$\frac{d^2P}{dx^2} = \frac{2}{3} > 0$$

So $x = 6$ minimises P. ✓

So $P = 2x + \frac{72}{x}$.

When $x = 6$,

$$P = 2(6) + \frac{72}{6}$$
$$= 24 \text{ metres.} \checkmark$$

Very familiar method applied here. Learn to differentiate fractions with x in the denominator by converting to negative indices. Students are advised to practise optimisation questions given how often they are included in HSC exams, often towards the end. Notice that the answer is given in part **a**. This is because, even if you don't answer part **a** successfully, you can still use this answer to complete part **b**. Also, check that you have actually answered the question. Did you find the minimum perimeter and state what it was?

Question 26 (7 marks)

a $A_0 = \$60\,000$

$A_1 = \$60\,000(1.005) - 800$
$= \$59\,500$ ✓

$A_2 = \$59\,500(1.005) - 800$
$= \$58\,997.50$

$A_3 = \$58\,997.50(1.005) - 800$
$= \$58\,492.4875$
$\approx \$58\,492.49$ ✓

Recurrence relations are specifically mentioned in the new course and they are common content with Maths Standard 2.

b As the balance in the account changes each month, so too does the amount of interest earned.

Although, not absolutely required, a table may be useful.

Month	Opening balance	Interest (0.5%)	Withdrawal	Closing balance
1	60 000	300	800	59 500
2	59 500	297.50	800	58 997.50
3	58 997.50	294.99	800	58 492.49 ✓
Total		892.49		

Therefore, \$892.49 in interest has been earned in the first 3 months. ✓

OR Decrease in balance = \$60 000 – \$58 492.49
= \$1507.51 ✓

Total withdrawals = 3 × \$800
= \$2400

Interest earned = \$2400 – \$1507.51
= \$892.49 ✓

The mathematics here does not extend beyond addition, subtraction and percentages. This question tests deep understanding of annuities. The 'trick' here is to realise that the interest earned each month is not constant. Practise solving these recurrence relations financial maths problems often so that by the time you see one in an exam, you know exactly what needs to be done. Don't waste time 'overthinking' and 'overwriting'.

c $A_1 = 60\,000(1.005) - 800$

$A_2 = A_1(1.005) - 800$
$= [60\,000(1.005) - 800](1.005) - 800$
$= 60\,000(1.005)^2 - 800(1.005) - 800$
$= 60\,000(1.005)^2 - 800(1 + 1.005)$

$A_3 = [60\,000(1.005)^2 - 800(1.005) - 800](1.005) - 800$
$= 60\,000(1.005)^3 - 800(1.005^2) - 800(1.005) - 800$
$= 60\,000(1.005)^3 - 800(1 + 1.005 + 1.005^2)$ ✓

Generalising the pattern:

$A_n = 60\,000(1.005^n) - 800(1 + 1.005 + 1.005^2 + \cdots + 1.005^{n-1})$

$= 60\,000(1.005^n) - 800\left(\frac{1.005^n - 1}{0.005}\right)$ ✓

$= 60\,000(1.005^n) - 160\,000(1.005^n - 1)$
$= 160\,000 - 100\,000(1.005^n)$

Therefore, when $n = 94$,

$A_{94} = 160\,000 - 100\,000(1.005^{94})$
$= \$187.85$ (nearest cent). ✓

The expression for A_n had to be established to find A_{94} directly. Again, this is a commonly asked series problem, so learn how to do this properly beforehand so that it becomes intuitive and straightforward during the exam. Notice how factorising the 800 comes in handy and that the 0.005 in the denominator divides into it easily. Calculator skills are also important here.

Question 27 (5 marks)

The median temperature is 22°C, which can be determined from the box-and-whisker plot.

So the mean temperature ($\overline{x}$) is 21.475 since it is 0.525 below the median. ✓

$\overline{y} = \frac{684}{20} = 34.2$ ✓

The equation of the median line of regression is given as $y = -10.6063 + bx$. Given that it passes through the point $(\overline{x}, \overline{y})$, we can use this to determine b.

$y = -10.6063 + bx$

Substitute (21.475, 34.2)

$$34.2 = -10.6063 + b(21.475)$$
$$21.475b = 44.8063$$
$$b = 2.08644\ldots$$ ✓

So $y = -10.6063 + 2.08644\ldots x$. ✓

When $x = 19$,

$y = -10.6063 + 2.08644\ldots(19)$
≈ 29 chirps per interval. ✓

The famous 'crickets' question of 2020 was one of those long 5-mark HSC exam questions with a lot of reading and thinking involved, and common content with Maths Standard 2. This is called an unscaffolded question because there aren't 'parts' that give you clues or steps towards the final answer, instead you're on your own. A straightforward question in terms of calculations and mathematical methods applied, but it has a high reliance on literacy and interpretation, raising the complexity. Spend time reading and highlighting key information, then thinking about possible strategies. Don't start writing straight away. Ask yourself: What is the question asking for? Then work towards a solution. A common student error was rounding partial answers too early, rather than at the end.

Question 28 (5 marks)

a Since the rates of pay are normally distributed with a mean of \$25 and a standard deviation of \$5, the representation of the distribution is shown below.

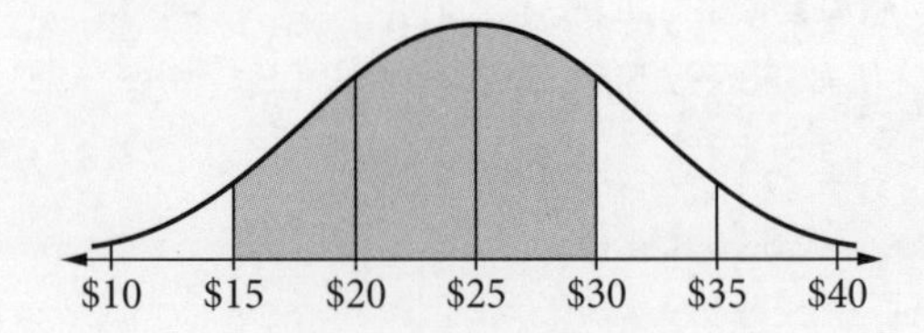

The percentage of scores in the range $-2 \le z \le 1$

$= \left(68 + \frac{95 - 68}{2}\right)\%$

$= 81.5\%$ ✓

So P(at least one earns between \$15 and \$30)

$= 1 - P$(neither earns between \$15 and \$30) ✓

$= 1 - (1 - 0.815)^2$

$= 1 - 0.185^2$

$= 0.965\,775$ ✓

Quite an involved question that ties z-scores with two-stage probability. Learn to work with the percentages of the empirical rule efficiently (these are on the HSC exam reference sheet). It's always good to draw a diagram, even if it's a rough one, to help you with the percentages. Also, remember that 'at least one' = 'not none'.

b Workers : non-workers = 3 : 1

So P(an adult works) $= \frac{3}{4}$ ✓

and $P(X \ge 25) = \frac{1}{2}$ as mean = 25.

P(adult earns more than \$25)

$= \frac{3}{4} \times \frac{1}{2}$

$= \frac{3}{8}$ ✓

Difficult, because it was unusual to see a probability expressed as a ratio. Also, some students confused two-stage probability with conditional probability and tried to use a formula instead.

Question 29 (4 marks)

a $y = c\ln x$

So $y' = \frac{c}{x}$.

When $x = p$,

$y' = \frac{c}{p}$ ✓

When $x = p$, $y = c\ln p$

so the equation of the tangent is:

$$y - c\ln p = \frac{c}{p}(x - p)$$
$$y - c\ln p = \frac{c}{p}x - c$$
$$y = \frac{c}{p}x - c + c\ln p \text{ ✓}$$

Although finding the equation of a tangent has been done many times by students, the general nature of the function in this question (all variables, no numbers) raises its complexity significantly. Some students did not treat c as a constant and tried to use the product rule.

b We require the gradient, $\frac{c}{p} = 1$. ✓

So $c = p$.

When $x = 0$, $y = 0$,

so $0 = 0 - c + c\ln p$

$c\ln c - c = 0$ (as $c = p$)

$\ln c - 1 = 0$ (as $c \neq 0$)

$\ln c = 1$

So $c = e$. ✓

Difficult question targeting the more capable students and related to part **a**. Learn to work with constants, such as c and p, treating them just like a number. Also, practise solving logarithmic equations, and noting that the solution is an exact value, e.

Question 30 (6 marks)

a Solving simultaneously,

$ax^2 = 4x - x^2$ (as the x coordinate of A is not 0) ✓

$ax = 4 - x$

$ax + x = 4$

$x(a + 1) = 4$

So $x = \frac{4}{a+1}$, as required. ✓

Make sure your algebra is strong. You can divide both sides of an equation by x if and only if you know that $x \neq 0$.

b Integrand $= 4x - x^2 - ax^2$
$= 4x - (1 + a)x^2$

$$\text{Area} = \int_0^{\frac{4}{a+1}} \left(4x - (1+a)x^2\right)dx \text{ ✓}$$

$$\frac{16}{3} = \left[2x^2 - \frac{(1+a)}{3}x^3\right]_0^{\frac{4}{a+1}} \text{ ✓}$$

$$\frac{16}{3} = 2\left(\frac{4}{a+1}\right)^2 - \left(\frac{1+a}{3}\right)\left(\frac{4}{a+1}\right)^3 - 0$$

$$\frac{16}{3} = \frac{32}{(a+1)^2} - \left(\frac{1+a}{3}\right)\frac{64}{(a+1)^3}$$

$$\frac{16}{3} = \frac{32}{(a+1)^2} - \frac{64}{3(a+1)^2}$$

$$16 = \frac{96}{(a+1)^2} - \frac{64}{(a+1)^2}$$

$$16 = \frac{32}{(a+1)^2} \text{ ✓}$$

$(a+1)^2 = 2$

$a + 1 = \sqrt{2}$ ($a > 0$, so $a + 1 > 0$)

So $a = \sqrt{2} - 1$ ✓

Complex HSC question targeting the more able students. Heavy reliance on well-developed algebraic skills being applied to not just equations but to integration. Students made many careless errors in algebra. Note the link to part **a**.

Question 31 (7 marks)

a

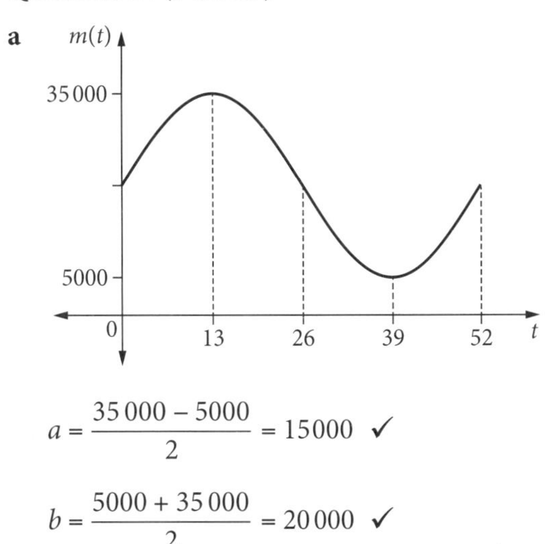

$$a = \frac{35\,000 - 5000}{2} = 15000 \text{ ✓}$$

$$b = \frac{5000 + 35\,000}{2} = 20\,000 \text{ ✓}$$

Straightforward question on translating a trigonometric function. Learn the properties of trigonometric functions, such as amplitude and centre, rather than memorising formulas.

b From the graph of $m(t)$, $m'(t) > 0$ for $0 \le t < 13$ and $39 < t \le 52$. ✓

Sketching $c(t)$:

Maximum value, which is 200, occurs when $\cos\left(\frac{\pi}{26}(t-10)\right) = -1$.

That is, when $\frac{\pi}{26}(t-10) = \pi$.

$$t - 10 = 26$$

$$\text{So } t = 36.$$

The period is 52, so the minimum value occurs at $36 - 26 = 10$, as 26 is half of 52.

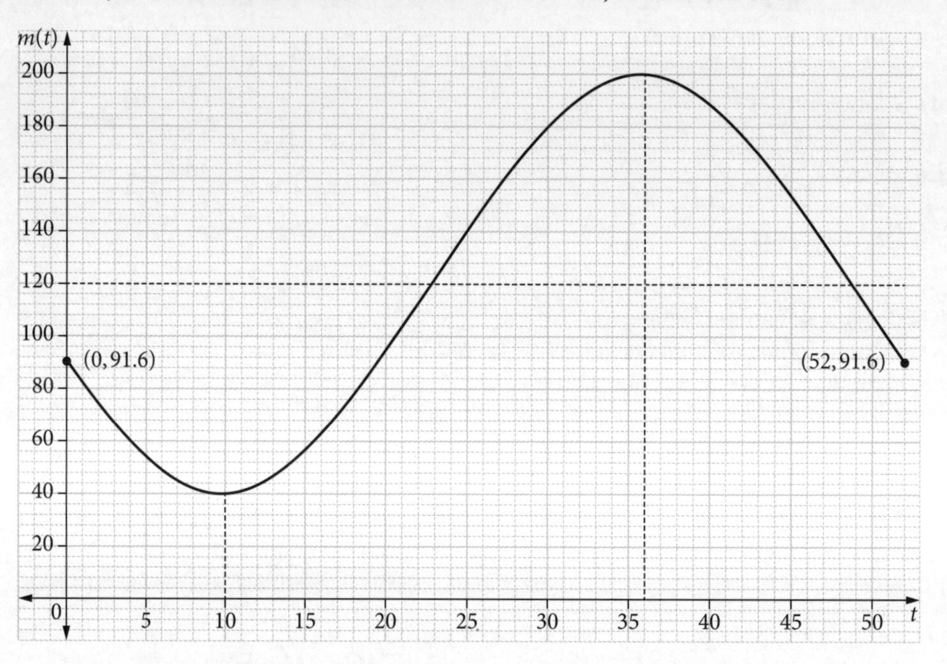

So $c(t)$ is increasing for $10 < t < 36$ ✓ and both will be increasing for $10 < t < 13$. ✓

This question is an application of a trigonometric function to population growth. The values of t for which $m(t)$ is increasing can be taken directly from the graph in part **a**. A sketch of $c(t)$ is the fastest way of determining the values of t for which $c(t)$ is increasing. But to do this, a deeper understanding of trigonometric graphs and their translations is required. Calculus could be applied, but it is more time-consuming and open to error. To achieve at a high level, you must *understand* Mathematics; not just *do* Mathematics. There is a big difference.

c Required to find $m'(36)$.

$$m(t) = 15\,000\sin\left(\frac{\pi}{26}t\right) + 20\,000$$

$$\text{So } m'(t) = \frac{7500\pi}{13}\cos\left(\frac{\pi}{26}t\right) \checkmark$$

$$m'(36) = \frac{7500\pi}{13}\cos\left(\frac{\pi}{26}(36)\right)$$

$$\approx -643 \text{ mice/week.} \checkmark$$

Straightforward process, but a complicated trigonometric function must be differentiated.
Note the links to parts **a** and **b**.

HSC exam reference sheet

Mathematics Advanced, Extension 1 and Extension 2

Note: Unlike the actual HSC reference sheet, this sheet indicates which formulas are Mathematics Extension 1 and 2.

Measurement

Length

$$l = \frac{\theta}{360} \times 2\pi r$$

Area

$$A = \frac{\theta}{360} \times \pi r^2$$

$$A = \frac{h}{2}(a + b)$$

Surface area

$$A = 2\pi r^2 + 2\pi rh$$

$$A = 4\pi r^2$$

Volume

$$V = \frac{1}{3}Ah$$

$$V = \frac{4}{3}\pi r^3$$

Financial Mathematics

$$A = P(1 + r)^n$$

Sequences and series

$$T_n = a + (n - 1)d$$

$$S_n = \frac{n}{2}[2a + (n - 1)d] = \frac{n}{2}(a + l)$$

$$T_n = ar^{n-1}$$

$$S_n = \frac{a(1 - r^n)}{1 - r} = \frac{a(r^n - 1)}{r - 1}, r \neq 1$$

$$S = \frac{a}{1 - r}, \ |r| < 1$$

Functions

$$x = \frac{-b \pm \sqrt{b^2 - 4ac}}{2a}$$

For $ax^3 + bx^2 + cx + d = 0$: * *EXT1

$$\alpha + \beta + \gamma = -\frac{b}{a}$$

$$\alpha\beta + \alpha\gamma + \beta\gamma = \frac{c}{a}$$

$$\text{and } \alpha\beta\gamma = -\frac{d}{a}$$

Relations

$$(x - h)^2 + (y - k)^2 = r^2$$

Logarithmic and Exponential Functions

$$\log_a a^x = x = a^{\log_a x}$$

$$\log_a x = \frac{\log_b x}{\log_b a}$$

$$a^x = e^{x \ln a}$$

Trigonometric Functions

$$\sin A = \frac{\text{opp}}{\text{hyp}},\ \cos A = \frac{\text{adj}}{\text{hyp}},\ \tan A = \frac{\text{opp}}{\text{adj}}$$

$$A = \frac{1}{2}ab\sin C$$

$$\frac{a}{\sin A} = \frac{b}{\sin B} = \frac{c}{\sin C}$$

$$c^2 = a^2 + b^2 - 2ab\cos C$$

$$\cos C = \frac{a^2 + b^2 - c^2}{2ab}$$

$$l = r\theta$$

$$A = \frac{1}{2}r^2\theta$$

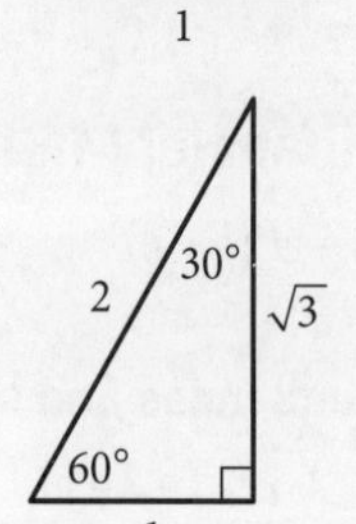

Trigonometric identities

$$\sec A = \frac{1}{\cos A},\ \cos A \neq 0$$

$$\operatorname{cosec} A = \frac{1}{\sin A},\ \sin A \neq 0$$

$$\cot A = \frac{\cos A}{\sin A},\ \sin A \neq 0$$

$$\cos^2 x + \sin^2 x = 1$$

Compound angles*

$$\sin(A + B) = \sin A\cos B + \cos A\sin B$$

$$\cos(A + B) = \cos A\cos B - \sin A\sin B$$

$$\tan(A + B) = \frac{\tan A + \tan B}{1 - \tan A\tan B}$$

If $t = \tan\frac{A}{2}$, then $\sin A = \frac{2t}{1 + t^2}$

$$\cos A = \frac{1 - t^2}{1 + t^2}$$

$$\tan A = \frac{2t}{1 - t^2}$$

$$\cos A\cos B = \frac{1}{2}\left[\cos(A - B) + \cos(A + B)\right]$$

$$\sin A\sin B = \frac{1}{2}\left[\cos(A - B) - \cos(A + B)\right]$$

$$\sin A\cos B = \frac{1}{2}\left[\sin(A + B) + \sin(A - B)\right]$$

$$\cos A\sin B = \frac{1}{2}\left[\sin(A + B) - \sin(A - B)\right]$$

$$\sin^2 nx = \frac{1}{2}(1 - \cos 2nx)$$

$$\cos^2 nx = \frac{1}{2}(1 + \cos 2nx)$$

Statistical Analysis

$$z = \frac{x - \mu}{\sigma}$$

An outlier is a score
less than $Q_1 - 1.5 \times \text{IQR}$
or
more than $Q_3 + 1.5 \times \text{IQR}$

Normal distribution

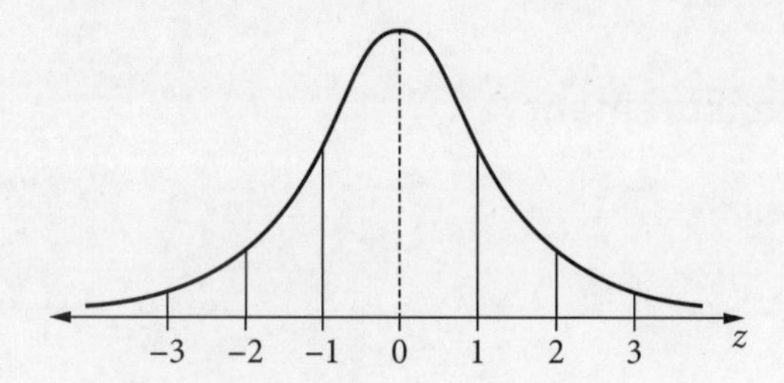

- approximately 68% of scores have z-scores between −1 and 1
- approximately 95% of scores have z-scores between −2 and 2
- approximately 99.7% of scores have z-scores between −3 and 3

Discrete random variables

$$E(X) = \mu$$

$$\text{Var}(X) = E\left[(X - \mu)^2\right] = E(X^2) - \mu^2$$

Probability

$$P(A \cap B) = P(A)P(B)$$

$$P(A \cup B) = P(A) + P(B) - P(A \cap B)$$

$$P(A|B) = \frac{P(A \cap B)}{P(B)},\ P(B) \neq 0$$

Continuous random variables

$$P(X \leq r) = \int_a^r f(x)\,dx$$

$$P(a < X < b) = \int_a^b f(x)\,dx$$

Binomial distribution*

$$P(X = r) = {}^nC_r p^r (1 - p)^{n-r}$$

$$X \sim \text{Bin}(n, p)$$

$$\Rightarrow P(X = x)$$

$$= \binom{n}{x} p^x (1 - p)^{n-x},\ x = 0, 1, \ldots, n$$

$$E(X) = np$$

$$\text{Var}(X) = np(1 - p)$$

*EXT1

Differential Calculus

Function	Derivative
$y = f(x)^n$	$\frac{dy}{dx} = nf'(x)\left[f(x)\right]^{n-1}$
$y = uv$	$\frac{dy}{dx} = u\frac{dv}{dx} + v\frac{du}{dx}$
$y = g(u)$ where $u = f(x)$	$\frac{dy}{dx} = \frac{dy}{du} \times \frac{du}{dx}$
$y = \frac{u}{v}$	$\frac{dy}{dx} = \frac{v\frac{du}{dx} - u\frac{dv}{dx}}{v^2}$
$y = \sin f(x)$	$\frac{dy}{dx} = f'(x)\cos f(x)$
$y = \cos f(x)$	$\frac{dy}{dx} = -f'(x)\sin f(x)$
$y = \tan f(x)$	$\frac{dy}{dx} = f'(x)\sec^2 f(x)$
$y = e^{f(x)}$	$\frac{dy}{dx} = f'(x)e^{f(x)}$
$y = \ln f(x)$	$\frac{dy}{dx} = \frac{f'(x)}{f(x)}$
$y = a^{f(x)}$	$\frac{dy}{dx} = (\ln a)f'(x)a^{f(x)}$
$y = \log_a f(x)$	$\frac{dy}{dx} = \frac{f'(x)}{(\ln a)f(x)}$
$y = \sin^{-1} f(x)$	$\frac{dy}{dx} = \frac{f'(x)}{\sqrt{1 - [f(x)]^2}}$ *
$y = \cos^{-1} f(x)$	$\frac{dy}{dx} = -\frac{f'(x)}{\sqrt{1 - [f(x)]^2}}$ *
$y = \tan^{-1} f(x)$	$\frac{dy}{dx} = \frac{f'(x)}{1 + [f(x)]^2}$ *

Integral Calculus

$$\int f'(x)[f(x)]^n dx = \frac{1}{n+1}[f(x)]^{n+1} + c$$
where $n \neq -1$

$$\int f'(x)\sin f(x)\,dx = -\cos f(x) + c$$

$$\int f'(x)\cos f(x)\,dx = \sin f(x) + c$$

$$\int f'(x)\sec^2 f(x)\,dx = \tan f(x) + c$$

$$\int f'(x)e^{f(x)}\,dx = e^{f(x)} + c$$

$$\int \frac{f'(x)}{f(x)}dx = \ln|f(x)| + c$$

$$\int f'(x)a^{f(x)}\,dx = \frac{a^{f(x)}}{\ln a} + c$$

$$\int \frac{f'(x)}{\sqrt{a^2 - [f(x)]^2}}\,dx = \sin^{-1}\frac{f(x)}{a} + c \text{ *}$$

$$\int \frac{f'(x)}{a^2 + [f(x)]^2}\,dx = \frac{1}{a}\tan^{-1}\frac{f(x)}{a} + c \text{ *}$$

$$\int u\frac{dv}{dx}dx = uv - \int v\frac{du}{dx}\,dx \text{ **}$$

$$\int_a^b f(x)\,dx$$
$$\approx \frac{b-a}{2n}\left\{f(a) + f(b) + 2\left[f(x_1) + \cdots + f(x_{n-1})\right]\right\}$$
where $a = x_0$ and $b = x_n$

*EXT1, **EXT2

Combinatorics*

$${}^nP_r = \frac{n!}{(n-r)!}$$

$$\binom{n}{r} = {}^nC_r = \frac{n!}{r!(n-r)!}$$

$$(x+a)^n = x^n + \binom{n}{1}x^{n-1}a + \cdots + \binom{n}{r}x^{n-r}a^r + \cdots + a^n$$

Vectors*

$$|\underset{\sim}{u}| = \left|x\underset{\sim}{i} + y\underset{\sim}{j}\right| = \sqrt{x^2 + y^2}$$

$$\underset{\sim}{u} \cdot \underset{\sim}{v} = |\underset{\sim}{u}||\underset{\sim}{v}|\cos\theta = x_1x_2 + y_1y_2,$$

where $\underset{\sim}{u} = x_1\underset{\sim}{i} + y_1\underset{\sim}{j}$

and $\underset{\sim}{v} = x_2\underset{\sim}{i} + y_2\underset{\sim}{j}$

$\underset{\sim}{r} = \underset{\sim}{a} + \lambda\underset{\sim}{b}$ **

Complex Numbers**

$$\begin{aligned} z = a + ib &= r(\cos\theta + i\sin\theta) \\ &= re^{i\theta} \end{aligned}$$

$$\begin{aligned} \left[r(\cos\theta + i\sin\theta)\right]^n &= r^n(\cos n\theta + i\sin n\theta) \\ &= r^n e^{in\theta} \end{aligned}$$

Mechanics**

$$\frac{d^2x}{dt^2} = \frac{dv}{dt} = v\frac{dv}{dx} = \frac{d}{dx}\left(\frac{1}{2}v^2\right)$$

$$x = a\cos(nt + \alpha) + c$$

$$x = a\sin(nt + \alpha) + c$$

$$\ddot{x} = -n^2(x - c)$$

*EXT1, **EXT2